PENGUIN BOOKS

HISS & TELL

Renowned feline behaviorist Pam Johnson-Bennett is the award-winning author of five books on cat behavior, including *Think Like a Cat* and *Twisted Whiskers.* Her first book was inducted into the *Cat Fancy* magazine Hall of Fame as one of the best books on cats ever published. Self-taught, she began her career when her own problem cat was labeled "hopeless" by the vet. Pam educated herself on animal psychology in order to save her cat's life. After successfully treating her own cat, as well as hundreds of other so-called "hopeless" pets, she became a veterinary technician while continuing to build her behavior consultation practice.

Considered one of this country's leading experts on cat behavior, Pam runs a private vet-referred counseling practice in Tennessee. She also writes a monthly behavior column for *Cats* magazine and is the cat expert at ivillage.com. Pam is a popular guest on national TV and radio as well as a contributing expert to all of the leading cat publications.

Pam lives in Nashville with her husband and three cats.

Pam Johnson-Bennett
FELINE BEHAVIORIST

Hiss & Tell

True Stories from the Files of a Cat Shrink

PENGUIN BOOKS

PENGUIN BOOKS
Published by the Penguin Group
Penguin Group (USA) Inc., 375 Hudson Street, New York, New York 10014, U.S.A.
Penguin Group (Canada), 10 Alcorn Avenue, Toronto,
Ontario, Canada M4V 3B2 (a division of Pearson Penguin Canada Inc.)
Penguin Books Ltd, 80 Strand, London WC2R 0RL, England
Penguin Ireland, 25 St Stephen's Green, Dublin 2, Ireland (a division of Penguin Books Ltd)
Penguin Group (Australia), 250 Camberwell Road, Camberwell,
Victoria 3124, Australia (a division of Pearson Australia Group Pty Ltd)
Penguin Books India Pvt Ltd, 11 Community Centre,
Panchsheel Park, New Delhi – 110 017, India
Penguin Group (NZ), cnr Airborne and Rosedale Roads,
Albany, Auckland, New Zealand (a division of Pearson New Zealand Ltd)
Penguin Books (South Africa) (Pty) Ltd, 24 Sturdee Avenue,
Rosebank, Johannesburg 2196, South Africa

Penguin Books Ltd, Registered Offices: 80 Strand, London WC2R 0RL, England

First published in the United States of America by
The Crossing Press, Inc. 1996
This revised edition published in Penguin Books 2001

11 13 15 17 19 20 18 16 14 12

LIBRARY OF CONGRESS CATALOGING IN PUBLICATION DATA
Johnson-Bennett, Pam, 1954–
Hiss and tell: true stories from the files of a cat shrink /
Pam Johnson-Bennett.
p. cm.
Includes bibliographical references.
ISBN 0 14 02.9853 3
1. Johnson-Bennett, Pam, 1954– 2. Veterinarians—Biography.
3. Cats—Behavior. 4. Cats—Psychology. I. Title.
SF613.J64 A3 2001
636.8'089689—dc21 00-040634

Printed in the United States of America
Set in Bodoni Book
Designed by Lorelle Graffeo

In loving memory of

my father, Albert Johnson, and

my precious cat, Olive.

Their brief time here made

the world much more beautiful.

Preface

I am a feline behaviorist—something my mother still has trouble admitting to people. It certainly is an unusual job, one that most people have never heard of, but it is also extremely rewarding. I'm self-taught, though, in my opinion, I've actually had the best teachers in the world—the cats themselves. I learned this job through years of hard work, research, trial, and a whole lot of error. Though there were many people who didn't think this would become a worthwhile career (actually, that's putting it mildly—most of them thought I was out of my mind), I knew there were cats out there who needed me.

I work on a veterinarian referral house-call basis. After the doctor examines the cat and determines that there is no medical cause for the undesirable or unusual behavior, he or she then gives the client my telephone number. My job is to visit the home, try to determine the cause of the problem, and set up a behavior-modification plan.

What I find so fascinating and challenging about my job is that, in order to understand the cat's behavior, I also have to look at the owner and how they both interact. The relationship between a cat and his human family is often the key to why the cat acts the way he does.

I've met many unforgettable cats. There have been a great many truly memorable owners, as well. When I set out on a house call, I never know what I'm going to find. It can be as simple as a cat urinating on the carpet because the litter box is just too dirty (these cases can be trickier than you'd think, since I'm not known for my diplomacy), or it can be as delicate a problem as helping an owner with a dangerously aggressive cat. The types of problems I see include stress, depression, aggression, boredom, grief, neglect, abuse, fear, competition, and just about anything else you can think of.

Over the years, my profession has expanded far beyond making behavior-counseling house calls. Owners have called me with all kinds of cat-related requests. I've taken difficult-to-handle cats to the vet for frightened owners, cried along with grieving family members as they've watched their beloved cats die, negotiated pet visiting rights for divorcing couples, and helped cat-hating new spouses adjust to life with a feline. I'm very often the first one called to help track down a lost cat or rescue a stray.

There are times when this job is a very sad one. I've walked into heartbreakingly abusive households, and, even though I've been able to remove the cats from the abusive environment, the memory of what they went through haunts me every day.

I've also been privileged to witness unconditional love, heartwarming human-animal friendships, and endless patience—often the owner's, but mostly the cat's.

Aggression is the second most common behavior problem I'm called about—litter box aversion being number one. A cat's aggression may be based on fear; it may be a response to predatory or territorial instincts; it may be food-related; or it may be related to medical problems. There are any number of reasons for aggressive behavior in a cat, but in the owner's mind it all comes down to one thing: the family pet has suddenly become an attack cat.

Often, when owners call me about an aggression problem, they're very embarrassed. They are humiliated to admit that they've suddenly become frightened of their little ten-pound kitty.

Feline aggression doesn't make newspaper headlines the way dog attacks do. Granted, the sight of a snarling sixty-pound dog sends shivers down my spine, but an aggressive cat is not to be taken lightly. A serious cat bite is very painful, can lead to infection, and can land you in the hospital. Scratches can also be very serious and often make you feel like you've just been attacked by a fur-covered razor blade.

People may snicker and make jokes about your cat being a rottweiler in disguise, but if you're dealing with an aggression problem, you need to discuss the situation with your vet. If, after doing a thorough exam on your cat, the vet determines there is no medical cause for the aggression, I suggest you seek out the advice of a behaviorist.

You see, very often I'm the last chance a cat gets. The number of animals who are euthanized for behavior problems is staggering. Many of the behavior problems originate from our thinking that cats are little fur-coated people. But as much as our feline friends become members of our family, we have a responsibility to meet their needs as *cats*. Too many people expect them to behave like perfect little children.

When we buy or adopt a puppy, one of the first purchases we make is a dog-training book. We also don't hesitate to seek advice from a vet or professional trainer. But when we buy or adopt a cat, we assume the cat comes pretrained. Cats are commonly described as "low-maintenance" pets. So when the cat misbehaves, we're convinced he's being mean or spiteful. It's not unusual for owners to say to me, "The cat *knows* he's being bad." That's not true. A cat doesn't deliberately set out to ruin our day by planning spiteful acts. He's reacting to a problem and trying to find a solution. Whatever is causing the cat to "misbehave" is also causing him a great deal of stress. Finding the *cause* of the problem is the key to success. Punishing the cat is not. Part of my clients' assignments involves looking at their home through the eyes of their cats. It's always an enlightening experience.

I love what I do for a living and wouldn't trade it for anything. Even though I've been scratched, bitten, swatted, hissed at, stalked, urinated on, ignored, run away from, growled at, and even vomited on, I still look forward to each day. Every time a cat and his owner reach an understanding, it's worth everything to me.

It wasn't my intention to write a book full of glowing tales of my successes, of how Pam Johnson-Bennett "saved the day." Rather, I've written this to give you a peek at some of the remarkable cats and owners who have left impressions on me. While you won't necessarily find common behavior problems here, I hope you'll enjoy learning about some of my more unusual house calls. Perhaps you might even acquire a new understanding of how cats view our world. If nothing else, in reading these stories, perhaps you'll realize that your own cat's behavior problems aren't so bad after all.

Pam Johnson-Bennett

Some names have been changed to protect the embarrassed.

Acknowledgments

Thank you to all of my clients (both feline and human) for your trust and confidence. I'm very lucky to be able to spend my days surrounded by cats and the people who love them.

Thank you to the veterinarians who have been so generous with your time over the years. A very special thanks to my friends Joe Ed Conn, D.V.M., and Mark Waldrop, D.V.M.

Thanks to Wendy Wolf and my family at Penguin.

I have been blessed with such wonderful family, friends, and colleagues who inspire, support, and love me even when my schedule causes me to be missing in action so many times. Thank you Mom, Kathren, Joan, Bud, Sharon, and Ginger. And thanks to my favorite writer, Chicago's pride and joy, Steve Dale.

To my agent, Linda Roghaar—you're the greatest!

Thank you to the most incredible man in the world—who, by the way, just happens to be my husband, Scott.

And finally, thank you to the best family of meows and bowwows: Albie, Bebe, Mary Margaret, and my dog, Annabelle, who is courageously battling cancer.

Contents

hiss & Tell

○●○

Tiki's Mysterious Mood

It was ten o'clock on Thursday night, and I had just gotten home from a late day of client appointments. My head throbbed, and my empty stomach was grumbling. I was looking forward to taking a hot shower, grabbing a little snack, and then settling into my comfortable bed.

It was another half-hour of checking my phone messages and returning several calls before I finally made it to the shower. As is usually the case, the telephone rang the moment I had a head full of lathered shampoo. Turning off the water, I stuck my head out of the shower to listen to the message.

"This is Margaret Taylor," a voice said urgently. "I've called everywhere. I don't know what to do. The animal hospital gave me your number. Please call me. My cat attacked me. I've locked myself in the bedroom. Call me right away." She left her phone number and again pleaded for me to call.

I hurried out of the shower and over to the answering machine to jot down the number. "No dinner tonight," I said to myself as I dialed the phone.

"Hello?" said a frightened voice.

"Hello, this is Pam Johnson-Bennett. Is this Margaret Taylor?" I could tell by her frantic tone that it was.

"Oh, thank goodness you called. I don't know what to do. Tiki's out in the hall growling," she said under her breath, and then she began to cry. "Please, can you come over?"

"First of all, are either you or the cat injured?"

"He bit me pretty hard, but I'm okay," she answered, trying to pull herself together. "I just can't believe it. Why is he acting this way?"

I asked Margaret Taylor to tell me exactly what happened.

Tiki, her four-year-old Siamese cat, had just launched a seemingly unprovoked attack on his owner. According to Mrs. Taylor, she was in the living room watching TV, when she heard a noise in the kitchen. Her husband was out of town on business, so she went to investigate. When she turned on the light, she saw Tiki crouched on the floor in the corner. The instant the cat saw her, he let out a screech and lunged at her, wrapping himself around her leg and biting into her calf. Mrs. Taylor screamed and had to forcibly pull Tiki off. He then bolted out of the kitchen.

Shaking and confused, Mrs. Taylor went to the bathroom to check her wounds. After cleaning one bite wound and several scratches, she looked around the house for Tiki, but couldn't find him anywhere. She decided to phone her husband and tell him what had just happened. Deciding to use the bedroom phone so she could sit on the bed, she stepped into the darkened room and flipped on the light switch. Again she heard that same screeching sound, and suddenly Tiki was charging at her again. Mrs. Taylor managed to pull the cat off her a second time and tossed him out into the hallway. After slamming the door, she ran to the phone to call her neighbor for help.

There was no answer, so Mrs. Taylor called her veterinarian. The after-hours recorded message referred her to the local animal emergency hospital. The doctor she then spoke to suggested she call me. By the time she called my number and heard my answering machine, she was in a panic, afraid there would be no one available to help her.

I told Mrs. Taylor I could be there in half an hour.

"There's just one thing," she began. "I'm afraid to leave the bedroom, so I won't be able to open the front door. My bedroom is on the first floor, and I'd climb out, but my house keys are in the front hall."

Uh-oh.

"Would you mind climbing in the bedroom window?"

I arrived at the Taylor home twenty minutes after our phone conversation. Before leaving the car, I reached into the glove compartment for my flashlight. The outside of the house was very dark. I wondered if any of the neighbors could see this strange figure inching around the dark house. As I turned the corner, I saw Mrs. Taylor's head peeking out the window. I walked closer and realized that, while she'd told me the truth about her bedroom being on the ground floor, she hadn't mentioned how high the window actually was.

"Mrs. Taylor," I called up to her, "I don't think I can reach this window without a ladder."

"My garage is locked," she answered in a loud whisper. "My neighbor never locks hers, though. She's not home, but you can go over there. I'm sure she must have a ladder." And she pointed to the house next door.

"Why don't I just try one of your other neighbors?" I suggested. Somehow the idea of poking around in somebody's garage at night didn't appeal to me.

"Are you crazy?" Mrs. Taylor shrieked. "Do you think I want all the neighbors knowing about this? Carol's the only one I trust!"

Again she pointed to the house next door—only this time with noticeably less patience.

So, off I went to search through one stranger's garage for a ladder so I could climb into the window of another stranger's house. Not the way I normally run my business. But then, I don't exactly have a normal business.

While I was unsuccessful in locating a ladder, I did find a stool. With that, I was able to hoist myself up (not a pretty sight) through the open window.

Once inside, the first order of business was to check Mrs. Taylor's injuries. The wounds weren't deep, but I advised her to have a doctor check them in the morning, just to be on the safe side.

"Mrs. Taylor, can you tell me what was going on right before the attack?"

She insisted that nothing unusual had happened before the incident. "The only difference was that Tiki wasn't on my lap in the living room the way he normally is when I watch TV," she sighed. "He always sleeps on my lap, and he did seem a little grumpy today."

After getting a little more history on the cat, it was time to meet Tiki and investigate. I explained to Mrs. Taylor that something may have agitated him earlier in the day and that he was probably still very aroused when she startled him in the kitchen.

"But we've lived here all of Tiki's life. Nothing's changed," she explained, a look of total confusion on her face.

"Something must have changed," I said while moving toward the bedroom door. "Now, I need you to stay calm so I can go out and see Tiki. I'd prefer you to stay in here."

"Are you going to try to capture him?" she asked.

"No, he's too upset. I just want to check on him."

Putting my hand on the doorknob, I turned and looked back at Mrs. Taylor. She was standing at the foot of her bed, chewing nervously on her nails. I noticed a small TV in the room. "Mrs. Taylor,"

I said, "turn the TV on and try to relax. You're safe in here." She nodded her head and obediently sat down in front of the TV.

I quietly opened the door and looked out into the hall. First, I wanted to restore some normalcy to the atmosphere of the house because, right now, both cat and owner were traumatized. Since Mrs. Taylor told me she watched TV every night, I turned on the set in the living room to reestablish that familiar sound for the cat.

I casually walked around the house, hoping to spot Tiki. As I was about to step into the hall, I caught a fleeting glimpse of black fur flying by into a small bedroom. I walked back to Mrs. Taylor's room with a question.

She jumped the moment I opened the door, then slowly sank back down onto the bed when she saw I didn't have her cat with me.

"You did say Tiki is a Siamese, right?" I asked.

"Oh, yes. Seal-point Siamese. Why?"

"And you did tell me there are no other pets in the house?" I glanced back toward the hall.

"Just Tiki. Why are you asking this?"

"I'll let you know in a minute," I said. "Just keep doing what you're doing."

I made my way back down the hall and toward the other bedroom. Tiki appeared at the far end of the hall, looking extremely tense. This was not the animal I had seen a minute ago.

"Hi, Tiki," I said softly. "I think I know what has you so upset."

Tiki's eyes followed me as I walked in, but he made no move to follow. I closed the door behind me. Kneeling by the bed, I lifted up the dust ruffle and saw a tiny pair of eyes staring back at me. This answered one mystery. I at least now knew what had set Tiki off. But where had this cat come from? And how could Mrs. Taylor not have known there was a strange animal in her house? I closed the mystery cat in the bedroom and went back to Mrs. Taylor's room.

"That's impossible," she stated after I informed her of my discovery. "Don't you think I'd notice another cat?"

"Well, yes, I would *think* so," I answered, truly puzzled. "But the evidence is sitting right under the bed in the next room."

Mrs. Taylor assured me there was no way a cat could get in her house, but I told her I was going to do a thorough check anyway. The cat must have gotten in somehow.

After checking all the doors and windows, I came across a small open window in the basement. That explained how the little visitor got in. Now, how did he get upstairs without detection?

When I reported the open window to Mrs. Taylor, she put her hands up to her face and said she had been painting a wooden rocking chair in the basement the day before. She had opened the window for fresh air and must have forgotten to close it again.

"My husband will have a fit when he finds out," she groaned.

I then asked her how the cat might have gotten upstairs. Did she leave the basement door open when she was painting?

"No," she said. "But the washer and dryer are downstairs. I always leave the door open when I do laundry so I can hear the buzzer on the dryer."

"You're lucky Tiki didn't get out through that open window," I said.

"Tiki never goes downstairs. He's afraid of the basement."

I went out to my car (via the front door this time) and got a carrying case for the black cat. As I walked down the front steps, my rumbling stomach reminded me of how hungry I was.

Back in the house, I stuck my head in the bedroom to tell Mrs. Taylor what I was doing. She was still too frightened to leave the room. This suited me just fine, since her fear certainly wasn't making things any easier.

Tiki was in the kitchen drinking from his water bowl. He looked more relaxed. I think he knew the situation was now under control. I put the carrying case down and walked toward the counter, talking to Tiki the whole time. I had seen a bag of cat food earlier and decided to give Tiki a snack as a distraction. As I filled the bowl, Tiki watched me without moving.

"You get an extra meal tonight," I said. Tiki waited until I was out of the room and then walked over to the bowl. One indication that an agitated cat has relaxed is when he resumes some of his normal activities, such as eating.

I approached the bedroom where the black cat was closed in and carefully opened the door—just in case he was planning to make a mad dash. But the cat was nowhere to be seen as I stepped in and quickly closed the door behind me. I knelt down next to the bed and peeked underneath. He was peeking right back at me.

"Are you hungry?" I asked as I pulled some of Tiki's cat food out of my pocket. I rolled a few pellets along the floor in the cat's direction. Quite hungry, the little visitor gobbled them up without hesitation. I rolled a few more pellets toward him. He then meowed softly, stretched, and walked out from under the bed. Before me sat a small black cat with beautiful green eyes. He was skinny, rather scruffy-looking, and had a torn left ear from some long-ago battle. He let out another quiet meow, and I offered him more food. When he was finished, he trotted over to me and hopped onto my lap. Stroking his back, I felt just how thin and matted he was. I also noticed that *he* was a *she*. "You lucked out, little girl," I said as I put her in the carrying case. "We'll find you a good home."

After depositing the carrying case in the front seat of my car, I walked back into the house and phoned a rescue volunteer I knew. I wanted her to come for the cat so I could work with Mrs. Taylor and Tiki. The volunteer said she'd get the cat from my car right away.

With the mystery now solved, I opened the door of Mrs. Taylor's bedroom.

"Let's go in the living room together and sit down," I said soothingly as I guided Mrs. Taylor out of the room. "I'd also like you to casually talk to Tiki while you're in there."

"Is he in there?" she asked, still worried.

"No," I said. "But wherever he is, I want the comforting sound of your voice to be available to him. He'll come out when he's ready."

"What'll I say?"

"Just talk to me about Tiki. Tell me about his good qualities. Talk about some of the wonderful memories you have," I instructed, hoping that remembering pleasant things would relax her. As a result, her voice would take on a more soothing quality, and this would benefit Tiki.

"Are you sure he won't attack me?" she asked, glancing toward the hall.

I had watched Tiki move around the house and was reasonably certain that Mrs. Taylor was in no danger. However, I did instruct her to keep the door of the extra bedroom closed until she could do a thorough cleaning to remove the scent of the black cat.

We discussed how the stray cat must have darted up the stairs, creating quite a scare for Tiki, who was faced with a little intruder in his domain. The attack on Mrs. Taylor occurred as a result of Tiki becoming startled when he was already in a highly agitated state. This behavior is known as redirected aggression.

We continued our conversation. I wanted Tiki to see that all was back to normal. It took a while, but Mrs. Taylor began to relax and, at one point, she even laughed at the chain of events that had occurred. Stifling a giggle, she informed me that I did, by the way, look pretty silly climbing through her bedroom window. Little did I know that entering clients' homes by means other than the usual front door would be something I'd do many times in my career.

Within a few minutes, Tiki sauntered into the living room and sat down a foot or two away from us. I told Mrs. Taylor to keep on talking; she was to let Tiki be the one to make the first move. It didn't take long before he jumped into her lap, curled up, and fell asleep.

Mrs. Taylor was given specific behavior-modification exercises to use with Tiki for a while. I explained that, even though he seemed to be back to normal, Tiki might still display some behavioral changes as a result of his traumatic experience. I wanted Mrs. Taylor to know exactly how to deal with Tiki, whatever he might need. I also asked if he was up-to-date on his vaccinations (which he was), since he'd had a close encounter with this visitor.

"What will happen to that black cat?" Mrs. Taylor asked as she walked me to the front door.

"She'll be checked over and tested for FELV/FIV," I said. "Then we'll get her spayed and placed in a foster home until she's adopted."

"Wait here," she said, then suddenly disappeared down the hall. A few minutes later she reappeared. "Put this toward that cat's future." She held out a handful of twenty-dollar bills.

When I got to my car, I looked at the money she'd given me. In addition to the check for my services, she had donated two hundred dollars toward her one-time houseguest.

As I started the car, I suddenly remembered that the stool was still sitting under the window. I was worried that Mrs. Taylor might have forgotten about it, and I knew she wouldn't want to explain to anyone why the stool was there. So I turned off the car, went around the back of the house, and returned the stool to her neighbor's garage.

During the drive home, I phoned the volunteer to check on the black cat. She reported that the cat was doing fine in the isolation room (a separate room for an incoming cat who hadn't been seen by the vet). The cat had eaten, used her litter box, and stretched out on the chair for a nap. Good, I thought. Now I can finally go home.

My follow-up calls to Mrs. Taylor over the next few weeks provided the good news that Tiki was doing wonderfully. I am happy to say that he has never displayed aggressive behavior since that night.

The future for the black cat turned out to be very bright. She got a good report from the vet, received vaccinations, and was spayed soon after. We found a wonderful home for her, and she now lives with two other cats. Missy, as her new family named her, turned out to be a beauty with a sleek, shiny coat to complement those gorgeous green eyes.

By the way, Mrs. Taylor informed me that her husband has since installed screens on the basement windows.

The Angel with Whiskers

My cat, Olive, frequently appeared on television with me. She was a medium-haired tortoiseshell tabby who came into my life several years ago in need of rescue. She had a perfect face for television and knew how to play it up for the camera. She often received more fan mail than I did.

It was after seeing Olive on TV several times that Elizabeth Donner called me with her unusual request.

"I know this is probably impossible, but my ten-year-old daughter would like to meet Olive," she began. "The reason I'm asking is that Nikki has leukemia. The chemo treatments leave her so debilitated that all she can do is stay in bed and watch television. She saw your cat one day and it perked her up so much. Nikki is crazy about cats and thinks Olive is very funny on TV."

There was a pause. I could hear her take in a very heavy breath. I was listening to a mother who was looking for whatever things in life could help get her child emotionally through another day, however small. Anything that could potentially bring a smile to her daughter's face became a mission.

"Mrs. Donner, I'm sure I can speak for Olive when I say that she'd be honored to visit your daughter," I answered, not even knowing where she lived or how I was going to do it.

A huge sigh was audible on the other end of the phone. "Oh, I'm so glad. I want this to be a surprise. Call me Liz, by the way."

I asked if they had any pets of their own. Liz answered that they had none. "I have to be honest," she said hesitantly, "I don't care for cats. Ever since I was scratched quite badly as a kid I've been wary of them. Nikki, my daughter, loves them, though. She has plenty of stuffed animals and she reads books about them."

We spoke for another twenty minutes as Liz told me more about Nikki, their only child. It was the story of a brave little girl who had spent the last two years of her short life in and out of hospitals. School was no longer a possibility during this past year as Nikki's good days became fewer and fewer.

A very popular child, Nikki had lots of friends who visited frequently up until the last few months. Even though she had never been afraid for her friends to see her bald head (she hated wearing scarves or hats), she now felt that her appearance made everyone uneasy. She was extremely pale with very dark circles that extended well beneath her eyes.

Liz wanted her daughter's friends to continue visiting and tried to keep Nikki's spirits up in every way possible. Nikki was adamant about keeping her friends away, though. She didn't want them to become afraid of her. She didn't think it was fair that they be subjected to seeing another child this way.

"She sounds like a truly courageous and remarkable person," I said as I thought back to my childhood and if I could've ever been that thoughtful if I'd been in Nikki's position. Probably not.

We made plans for me to bring Olive the following Saturday

when Liz's husband, Terry, could also be there. I hoped Olive would handle this well and be on her best behavior during what would certainly be an emotional day.

I was very honored that Nikki wanted to meet my cat but to be honest, it truly surprised me. Olive was very pretty and well behaved on camera (though there had been a few less-than-impressive moments when she let the cameraman know that he had gotten too close). But when I think of all the pet celebrities on TV, I doubted that Olive could even crack the top 500 list. Regardless, if Nikki wanted to meet Olive, then we were going to make that happen.

On that Saturday, Olive and I set out on our journey. Luckily, the Donners didn't live too far away. We pulled into the Donners' driveway three hours after we'd started out.

"Be on your best behavior today," I whispered to Olive as I reached over and pulled her carrier toward me. We looked at each other. "Do a good job in there and later, when we get back home I'll cook up a special treat for you." Olive watched me with interest. "You want chicken?" Olive just stared. "Turkey?" No response. "How about salmon?" At that, she gave me a slow blink as her tongue darted out to lick the tip of her nose. "Salmon it is, then," I declared with a wink as I reached across the seat to unhook the seat belt that I'd fasten around her carrier.

Liz and her husband Terry greeted me at the door, repeating "thank you" over and over. Though they both had big smiles on their faces, the matching dark circles under their eyes betrayed that there were many sleepless nights in this home. Liz, who had told me on the phone that she was thirty, looked at least ten years older. She was blond with fair skin and green eyes. Her long hair was pulled tightly into a ponytail, her bangs held back with a headband. When she reached out to shake my hand, I noticed her short, bitten fingernails. The gold wedding band and diamond engagement ring on her left finger clinked together as she moved her hands. They looked a size too large for her thin fingers. I guessed that she must've lost a significant amount of weight during this family crisis. I didn't know Terry's age but I was fairly certain that he was much younger than he looked. Nikki's cancer was stealing the lives of everyone in the Donner family.

"Is that Olive?" asked Terry as he pointed to the carrier at my side. "Would you like me to help you with that?" He reached his arm out toward the carrier but Liz put her hand out to stop him.

"Are you sure she won't scratch Nikki?" she asked.

I reassured her that I would hold Olive and not allow her near Nikki if I sensed any nervousness.

"Would you like something to drink? I can fix you some lunch if you're hungry," Liz offered, while her eyes never left the cat carrier.

"No, thank you," I answered, "let's go surprise Nikki." I knew that Liz wasn't comfortable around cats and there was no need to make her any more nervous than she already appeared.

Liz and Terry led the way down the hall toward Nikki's room. The door was partially closed. As we drew closer, I noticed Terry reach over and clasp Liz's hand in his.

Just before she opened the door, Liz glanced back at me. "Ready?" she asked. I nodded, noticed that her dark circles seemed to fade just as a little sparkle came into her eyes.

I quietly set the carrier down in the hallway, opened the door, and out walked Olive. Reaching down, I scooped her up in my arms. "Please," I whispered in her ear while digging in my pocket for a treat, "make me proud."

Terry pushed the door open, entering Nikki's room with Liz at his side. I stood in the hall, just outside of view, waiting for my cue.

"Nikki, honey," Liz said to her daughter, "do you feel up to a little company?"

"Who is it?" came the little girl's faint reply.

Terry answered. "Somebody who heard that you wanted to meet her. She was so excited that she drove all the way from Nashville."

"Nashville?" Nikki asked excitedly and in that moment I

thought, oh no, what if she's expecting it to be a country music star? Imagine, expecting Reba McEntire and getting Olive the cat instead.

"Are you ready, honey?" Liz asked as she walked back toward the doorway and motioned for me to enter.

I was barely in the room before I heard Nikki cry out "OLIVE!" At that moment I don't think Reba could've gotten a bigger smile out of that little girl.

Holding her arms out, Nikki asked if she could hold her, and Olive, who loved to be lavished with attention, was more than happy to oblige.

"How did you know?" Nikki asked, all smiles, as she petted Olive's head. I told her how her mother had contacted me.

I watched my cat carefully for a couple of minutes to make sure she was handling all of this without stress, and then allowed myself to just take in the whole scene. Here was a room that looked like the typical little girl's room—pink this, pink that, frills everywhere, stuffed animals, dolls, cat posters. Yet, in bed sat a little girl so atypical, a little girl forced to face things with the kind of strength that most adults don't even have.

Nikki herself was tiny. She seemed to practically disappear in the big bed. Because she was always cold, there were piles of blankets on the bed, making Nikki look even smaller.

Nikki's visit that day with my surprisingly well behaved, formerly tough-as-nails stray cat meant so much to everyone in that room. As we all struggled to hold back tears, Olive remained the picture of good manners and grace.

I brought Olive to visit Nikki again two weeks later and it was during that visit that Terry approached me with a request. Since Nikki loved Olive so much, he wanted to know if I would consider giving the cat to his daughter. Before I could even answer, Liz spoke up with her immediate objection.

"You know I don't like cats. We can't have a cat in this house, we just can't," she stated.

"It's for Nikki," Terry said softly.

Liz closed her eyes and slowly shook her head. "I can't take care of a cat on top of everything else, Terry. It's too much."

"I'll help. It'll mean so much to Nik," Terry pleaded, as they seemed to stare each other down.

Liz's expression slowly softened. "Okay," she finally said with a very tentative smile.

During this exchange, no one noticed that I had been attempting to answer Terry's question. I couldn't give Olive up. "She's my family," I replied. "Surely you must understand that."

Terry looked down at the floor and nodded, while Liz looked relieved.

"I do think that getting her a cat is a great idea, though," I added. "It doesn't have to be Olive. She'll love another cat. She just needs the love and companionship that the cat can offer."

Liz's expression became concerned again.

"Can it at least look like Olive?" Terry asked.

"Sure," I answered.

Liz interrupted. "Suppose she doesn't like the other cat?"

"Will you find the right cat for Nikki?" Terry asked as he looked first to me and then to his wife for approval. "Liz, is that okay with you?"

Liz stared at her husband for a long time, then looked back toward Nikki's door. When she faced up again, there were tears in her eyes. "If it'll help Nikki, I'll do it," she answered.

So, off I went in search of an Olive look-alike. Not such an easy task when you consider that the cat also needed to have a personality that would match Nikki's expectations.

Two weeks went by. I searched through shelters, answered countless newspaper ads, and went to every cat contact I had. Olive was apparently more unusual than I'd thought. Then, one evening a former client phoned me to ask if I could help find her cat another home. She was an older woman, a widow, with severe arthritis that was causing her increasing pain. Her inability to get around meant she was going to have to move in with her son and his family. Apparently, the conditions set by her son included no pets of any kind due to his severe allergies. It was hopeless trying to talk him into it, she'd said, so was there any way I could help her?

I don't usually agree to place cats, preferring instead to refer clients to the people I know who specialize in that exclusively. As the woman spoke, I went into my files and checked on the cat's history and physical description. I had originally been called to her house to help with another cat that had since died. The three-year-old cat she wanted to place didn't have any behavior problems. The physical description I'd written in my file didn't match Olive, but I agreed to help her anyway.

If you closed one eye and squinted with the other, the cat couldn't even pass for Olive in a pitch-black room. What impressed me, though, was her personality. She was sweet, affectionate, tolerant, and sociable. She also loved children. Mrs. Casey's neighbor had six kids (whew!) and her cat, Angel, seemed to enjoy every one of them. When asked if the neighbor wanted the cat, though, Mrs. Casey was given a polite but very firm "No!"

It was a long shot, but I brought Angel to Nikki to see if there was any chemistry. To my delight (sort of), she thought Angel was even better than Olive. So now Nikki had her new friend.

I helped the Donners get set up for Angel and provided them with several lessons in Cat Ownership 101. I brought along Angel's two scratching posts, litter box, litter, toys, and a supply of food. As my gift, I included two cat trees (one for Nikki's room). I wasn't doing it to be especially generous; my honest motivation was to ensure that Angel would prefer perching on the cat trees as opposed to the furniture, which would give Liz less reason to dislike her. Angel had been updated on vaccinations and all that was left for her to do was to settle in as a member of the Donner family.

And settle in she did! By the second evening, Angel became

Nikki's constant companion. They slept together, ate together, and when Nikki was well enough, played together. Angel put the much-needed spark back into Nikki's life. And even though Liz frequently grumbled about the cat hair and occasional food thefts perpetrated by Angel, she was thrilled to see Nikki so happy. As long as the cat kept her distance from Liz, she was satisfied. Angel seemed to sense this and stayed well out of her way.

Over the next four months, Nikki and I spoke on the phone several times. She always referred to Angel as her best friend. By the time eight months had passed, though, the calls stopped, as it required too much of an effort for Nikki to speak for any length of time. Then one evening, nine months after my first visit to the Donner house, I received a call from one of their neighbors. Nikki had died that morning.

Three weeks after Nikki's death, Liz phoned, pleading with me to come and take Angel.

"Why?" I asked.

"She's driving me crazy," Liz responded. "She meows all of the time, paces relentlessly, and yesterday she urinated on Nikki's bed. I can't deal with her anymore."

"Liz," I said quietly, "she misses Nikki. Just give her some time." There was a cold silence on the other end as I continued. "She doesn't understand what's happened. Her urinating on the bed is her way of leaving a scent for Nikki. It's her attempt to help Nikki find her way back to the nest," I waited. Silence. Then finally, Liz spoke.

"Nikki's dead. How do you explain that to the dumb cat?" she asked.

"What Angel needs is your attention. Play with her, talk to her. She's grieving."

"I don't like cats," Liz said, with strong emphasis on the word *cats*. "You knew that from the beginning."

I offered to help her work with Angel. I thought it might help with her own grief, but she was adamant about the fact that the cat must go. “Besides,” she said, “every time I see the cat it makes me miss Nikki even more.”

Liz’s husband didn’t agree with her but was concerned about how much stress she could handle right now. I reluctantly agreed to drive up and bring the cat back to Nashville. I planned to keep her myself this time and not attempt to place her. I figured she’d lost enough important people in her life and shouldn’t have to endure being tossed around again.

The weekend that I planned to go get Angel, the weather turned horrible in Nashville. The roads were icy and the forecast called for snow on top of that, which would make driving treacherous. I called the Donners to postpone our plans. Terry answered the phone and updated me on things.

“Liz hates the cat. She says that Angel’s always in the way and knocking things over. She’s terrified that the cat’s going to scratch her. The poor cat scrambles just to stay out of Liz’s path.”

“I’ll get there as soon as I can,” I said sadly.

“I appreciate whatever you can do,” Terry replied and we hung up.

Liz did her best to avoid the little cat as much as possible. She made sure the cat was fed, had fresh water and a clean litter box, but that was all. The only affection Angel received was from Terry. Liz refused to have any contact with the cat.

Angel was strictly forbidden in the Donners’ bedroom at night. The door to Nikki’s room was kept closed, and Angel basically lived on her cat tree in the den. Isolated and confused, she ate just enough food to survive and quietly made her way into the laundry room to use the litter box, careful to avoid coming in contact with Liz. Angel, Liz, and Terry were all living very separate, lonely lives.

On Monday, Terry had to fly out of town on business. Though very concerned about leaving his grieving wife alone, she convinced him that he needed to go. He had been away from work too long. His coworkers had been handling much of his duties, but he needed to be at this meeting in Chicago himself. I was planning to get Angel the next morning.

Alone in the house that evening, Liz tried to keep her mind on other things. She watched television, barely able to follow the plots of the programs. Eventually, she broke down in tears. Since Terry wasn't around, there was no need to pretend to be strong. Her tears turned to sobs until, finally exhausted, Liz fell asleep in the chair.

Angel watched the sleeping Liz from the top of her cat tree. Slowly and quietly, she made her way down the tree and gingerly walked across the floor. Looking up at Liz, Angel must've thought better of jumping into her lap, choosing instead to attempt a leap onto the end table. Unfortunately, though, Angel misjudged her ability to land on the slippery glass-top table and while struggling to maintain her footing, she sent the lamp crashing to the floor.

"Damn cat," screamed a startled Liz as she woke up and saw the shattered remains of the lamp on the floor, but Angel was already scurrying out of the room in terror. Liz followed her, fed up with the miserable pet.

Liz found Angel in the front hallway, looking toward the door. "You want to go out?" Liz asked sarcastically. "Well, why don't you just get the hell out of here?" With that, Liz swung open the door and the terrified cat was gone in a flash into the cold, dark night. Liz slammed the door.

After cleaning up the mess from the broken lamp and trying to watch a little more television, Liz sat at her desk in the den, planning to go through the mail. She'd been neglecting it for so long, and the pile was getting overwhelming.

She opened the top drawer and felt around for the letter opener. Rummaging through the clutter of junk, she stretched her fingers out, feeling for the thin plastic opener. Liz gave the drawer one good yank and found the letter opener wedged way in the back. Just before closing the drawer, she caught sight of a pile of pictures partially hidden in the corner. Carefully, she lifted up the stack of pictures and began looking at them. They were pictures that Terry had taken of Nikki during the last few months of her life. He must've hidden them in the drawer, Liz thought, afraid that it would be too difficult to look at them.

Liz reached over the pile of mail and turned on the desk lamp. Pushing aside the bills and condolence cards, she began to examine each picture of her precious daughter. In almost every one, Nikki was clutching her beloved cat, Angel. The smile on Nikki's face was illuminating. Liz noticed that except for Nikki's bald head, she looked like a normal, happy little girl with her pet. There were pictures of Nikki dressing a tolerant Angel up in doll clothes. There was a picture of Nikki pretending to have a tea party with Angel. The cat was wearing a little hat that Nikki and Liz had created out of paper. Liz couldn't help but chuckle at the silly expression on the cat's face.

As tears started burning in her eyes, she wiped them away. She wanted to see every picture clearly. She wanted to remember her child smiling and happy.

Finally, there were a few pictures of Nikki sleeping peacefully with Angel wrapped in her arms.

The pictures were reminding Liz of how much joy and love that little cat had brought to Nikki's life. She remembered how Angel barely left Nikki's side the night before she died. Liz couldn't get her daughter or the cat to eat that night and the only time she'd seen Angel leave Nikki's room was to use the litter box.

Suddenly, Liz turned and looked at the front door. "Oh, please be okay," she cried as she ran to the door and opened it wide. "Angel, Angel," she called out into the night. "Angel," she called again. No response.

Liz ran back into the house, looked up my number, and when I answered the phone, frantically told me what she'd done. "What should I do? How can I find her?" she cried.

"Very often, when an indoor cat ends up outside, they look for the closest cover they can find for safety. Look in the bushes just off the front porch or under the car," I advised her. "But Liz, don't run or yell because you'll only scare her even more. Call to her very quietly."

Going back outside, armed with a flashlight, Liz looked underneath the bushes near the porch. Angel was nowhere to be seen. "Angel," Liz called softly, trying not to cry and feeling so guilty about the way she'd treated the little cat.

Scanning the front yard with the flashlight, Liz felt hopeless about finding Angel. After all, why in the world would she want to come to someone who had treated her so badly? Liz knelt down in the grass. Suddenly, two little eyes appeared in the darkness. Angel was staring at her from deep inside an azalea bush. Liz's heart was pounding in her chest. She was ashamed of how cruelly she had treated her daughter's cat.

Very slowly, Liz walked over to the bush, and putting aside her lifelong fear of cats, reached underneath, gently grasping Angel. Too terrified to resist, the cat allowed herself to be pulled from the bush.

As she carried Angel into the house, Liz said, "I guess it's about time you and I get to know each other." With the confused cat in her arms, Liz phoned me to inform me that I wouldn't need to come for the cat the next day. "Angel's staying here," she said.

Angel and Liz slowly began developing their relationship. I conducted many counseling sessions with Liz by phone in order to help her become a more knowledgeable and tolerant cat owner. When I spoke to Terry a month later, he informed me that Angel had returned to her sociable, happy self and followed Liz all around the house.

Liz had come a long way in her attitude toward cats. The only rule she insisted on was that their master bedroom remain off-limits. Angel, grateful for the major strides her owner had made, obeyed by never setting foot in the forbidden area.

Quite unexpectedly, four months after Nikki's death, Liz was diagnosed with breast cancer. Due to her constant care of Nikki, Liz had neglected to have a small lump checked by the doctor. As a result, when she finally did get examined, surgery was immediately scheduled.

Back at home, while lying in bed, Liz, who had not even had sufficient time to grieve the loss of her daughter, now had to face her own uncertain future.

Waking from a restless afternoon nap, Liz sat up in bed and looked out the window. Terry was at work, so Liz was by herself. Her friends, though supportive, appeared awkward around her. After all, it must be difficult to watch one family endure so much sadness. And even though she knew Terry loved her, he was coping with his own pain and fear. Liz felt all alone.

Before settling back down to sleep, Liz happened to glance over to the doorway and caught sight of Angel sitting in the hall. Ever since Liz had returned from the hospital, Angel dutifully slept at the edge of the doorway, not daring to enter the room.

Liz patted the bed next to her. "You can come in, Angel."

The cat didn't budge. She knew the rules.

"It's okay, really, come in."

Not a whisker moved.

"Angel, I need you."

As if understanding the meaning of those words, the little cat tiptoed into the room, hopped on the bed, and curled up next to Liz.

When Terry came home that evening he found his wife sleeping peacefully with her arms wrapped around her little Angel.

Liz has remained cancer-free for the past four years and Angel sleeps right next to her every night.

○●○

The Paper Caper

It seemed that two-year-old Penelope, a black cat with white socks and one white ear, was obsessed with paper. When she was a kitten, her owners, the Bouchards, would crumple up pieces of paper for her to bat around, and it soon became a favorite game. The paper always ended up in shreds, but, being tolerant and loving owners, the Bouchards continued to indulge Penelope in this paper chase. Once it was over, Marianne Bouchard would simply use the vacuum to get rid of the evidence, while a satisfied Penelope lounged nearby. But apparently Penelope took this game more seriously than anyone thought.

The problem began when Penelope discovered that she didn't have to wait for her owners to provide the paper. There were several self-serve dispensers all around the house—rolls of toilet paper in both bathrooms, countless boxes of pop-up tissues, and the paper towels in the kitchen. Coming home from work one day, Marianne walked into a house covered in white confetti. Penelope had grabbed the end of a roll and charged throughout the house. She apparently enjoyed this experience so much that she repeated it with every roll.

Not knowing what else to do, the Bouchards were forced to keep everything out of Penelope's reach. All rolls of paper had to be kept in cabinets. Not very convenient, but at least it stopped Penelope—temporarily.

Now that her paper supply was cut off, the Bouchards were sure that life with Penelope would return to normal. But Penelope had other ideas. She was a cat on a mission. No paper was safe. The computer paper running through the printer was transformed into ribbons. Newspaper was shredded and chewed. Mail sitting on the desk was routinely nibbled. Keith Bouchard could not leave these paper items unattended for even one minute, lest the stealthy Penelope take advantage of the opportunity and pounce.

Then one Sunday morning, Penelope chewed up Keith's checkbook, and the Bouchards finally reached the limits of their patience. Keith insisted that Penelope be placed in another home, or else kept exclusively outdoors. Because of Penelope's paper obsession, Keith wasn't optimistic about finding the cat a home. Neither option was acceptable to Marianne. Tension between the couple grew.

The next day, a tearful Marianne phoned Penelope's veterinarian. He told her to bring the cat in for a complete examination, in case there was a medical reason causing her unusual behavior. He also told Marianne about me.

The results of Penelope's laboratory tests were all in the normal range, as were the findings of her physical exam. Marianne was both relieved and disappointed. She had convinced herself that there was surely a medical reason for the misbehavior and was confident the vet would be able to provide a solution.

The vet strongly recommended that the Bouchards contact me. When Marianne hung up the phone, she looked over at Keith, who'd been listening in on the cordless extension.

"We can't get rid of her," Keith said as he gently stroked the contented cat sleeping on his lap.

"I know." Marianne held up the slip of paper in her hand. "I guess I'd better call Pam Johnson-Bennett before Penelope eats this telephone number."

An appointment was scheduled for the next weekend.

It was a beautiful spring day when I drove to the Bouchard home, about twelve miles from my house. I rolled down my car windows and enjoyed the ride, feeling eager for summer to begin.

Seconds after I knocked, the door was opened by Marianne. "Thank you for coming," she said as she offered me her hand. "I've never met a behaviorist before. I hope you can help our cat."

Keith Bouchard was waiting for us in the den. He rose when we walked in and said how anxious he was to get down to work.

Keith was cooperative but nervous throughout the history-taking portion of our session. He was anxious for me to meet the paper-nibbling Penelope.

"I'll get her," Marianne volunteered. "She's always running in and out, but she comes whenever I call her."

When she returned to the room several minutes later, it was without the cat. "I can't find her anywhere," she said to her husband.

"Did you check outside?" he asked, and Marianne nodded.

The three of us began a search of the house, but with no success. We looked outside, but still no sign of Penelope. Marianne was becoming extremely concerned.

"Maybe she'll come out if she sees the toys I have," I said. I went back in the house to get "Da Bird," one of my favorite interactive toys, which has feathers on the end of a cord that hangs from a long wand, and simulates a flying bird. The three of us then sat on the front lawn while I showed the Bouchards how to use the toy. Marianne began waving it back and forth, all the while softly calling Penelope's name, but her beloved cat did not materialize.

It was getting late. I had already been there an hour and I needed to leave soon for my next appointment. "Based on what you've told me about Penelope, I can give you some behavior-modification techniques that I'm confident will help your situation," I said as we walked back into the house. "It won't be the first time that I haven't actually seen a client's cat during the house call."

I explained to the Bouchards that I felt Penelope's problem stemmed from boredom. Both owners worked two jobs and hadn't played with her in at least a year. This energetic cat spent all day and most of the night alone.

To remedy Penelope's boredom, I suggested the Bouchards consider adopting a second cat. I also recommended regular play sessions for Penelope with interactive toys. "At least two sessions a day, for fifteen minutes each."

Keith looked at me in disbelief. "Where am I going to find an extra thirty minutes a day to play with my cat?"

"Substitute the half-hour each day you used to spend cleaning up the paper she shredded," I offered.

I further instructed the Bouchards to leave out a few pieces of paper that had been coated with bitter apple. This liquid has a foul, bitter taste, and I knew Penelope would soon learn to connect the unpleasant taste with the paper.

After some final instructions about additional behavior-modification exercises, I promised the Bouchards I would call them in a week to check on their progress and offered to help them introduce a second cat to Penelope.

I was about to get in my car when I happened to look in through the open window. There, sitting so innocently on the front seat, was a white cat. "You must be Penelope."

Upon hearing the door open, Penelope got up, stretched, and gave a lazy yawn. I leaned toward the back seat to put my briefcase away, and when I straightened up again, Penelope was gone. Looking over my shoulder, I saw her strolling up the driveway toward the house. Midway, she stopped and turned to face me. It was then that I noticed she had something in her mouth. I strained to get a better look and then realized that the small white rectangle in her mouth was one of my business cards. She must have found it on one of the seats.

"Call me anytime," I said as I got in my car and waved good-bye.

○ ● ○

Winston's Toy

Barbara McMillan phoned to request my help when her two-year-old male cat stubbornly refused to accept the newest addition to the McMillan household—a fourteen-week-old female kitten. More than a month had passed, and things were only getting worse. Barbara informed me that she had conducted a proper introduction, but without success.

"My friend told us how to introduce cats according to the instructions in your book, and Winston still hates Courtney," she complained. "Are you sure your methods work?"

"Yes, Mrs. McMillan," I assured her, "they really do. But every cat has an individual personality, and some require a more customized approach. Even so, it's not uncommon for cats to take more than a month to become friends."

"I understand that," she said, "but Andy and I are quite upset about this. We'd like you to come to the house as soon as you can."

I agreed to come over the next evening.

Barbara and Andy McMillan greeted me at the door. Winston, a handsome, muscular, charcoal-gray cat, stood just behind them. He eyed me, suspicious but curious, as I walked through the door with my briefcase and an armful of assorted toys.

The McMillans led me to the living room. Barbara and her husband sat across from me on the sofa, grasping each other's hands. This was obviously a stressful occasion for both of them.

"I followed your instructions," Barbara explained after she'd finished giving me Winston's history. "Courtney's been in the guest bedroom since we first brought her home."

"She followed all the steps you recommend," Andy added, coming to his wife's defense.

They both looked at me as if to say they'd done their job, now what was I going to do about it?

"Oh, I believe you," I quickly said. "But, just to be thorough, let's review how you proceeded with the introduction."

Barbara related her beginning steps to me, and, sure enough, she had followed my guidelines. They brought the kitten in the house and put her in the guest bedroom—to prevent Winston from detecting the little intruder's scent directly on his owners. Courtney stayed in the bedroom for a week while both cats adjusted. Barbara said everything was going smoothly—Winston didn't seem at all disturbed that there was a stranger on the other side of the door. The kitten was adjusting to her new home as well. So far, so good, the McMillans thought. They began to do cat/room switches for short periods (that is, placing Winston in the bedroom and letting Courtney out so they would become familiar with each other's scent). Andy said the kitten loved being free to explore the house and that Winston wasn't bothered by the strange scents in the kitten's bedroom.

"He checked everything out, sniffed Courtney's litter box, and then fell asleep on the bed. Basically, it was very uneventful," Andy said, looking over at his wife.

"That's right," Barbara added. "We even thanked our friend Lucy for telling us about your method, because it seemed to be making the situation so much easier. We'd heard such horror stories about how mean cats can be when their owners bring another cat home. We thought we were pretty lucky."

There was a brief, uncomfortable silence as the two of them looked at me. I got the feeling they were waiting for me to admit my failure and be done with it.

"When did the trouble begin?" I asked.

"From the very moment they met face-to-face," Andy answered. "Every time since, Winston races over to Courtney and attacks her the second he sees her."

"When he spots Courtney, he's like a torpedo," Barbara agreed.

Winston sauntered into the room, and I slid down onto the floor to be at his level. He eyed the toys by the side of my chair with interest. When I picked one up and started wiggling it along the carpet, Winston responded by crouching down, his tail twitching in anticipation.

"What's that?" Barbara asked.

"It's called the Play-n-Squeak," I replied. It's a lightweight wand with a small mouse toy dangling from the end of a line. "Didn't the person who told you about my introduction method tell you about it? Toys are an important part of behavior modification."

"No, Lucy didn't mention any toys," Andy said. "She just told us the part about how to introduce the cats. Actually, Lucy wasn't the one who read your book," he admitted. "She said a friend of hers had read it."

"But anyway," Barbara was quick to add, "Winston has always had his own special toy. It's the only one he likes, so we just use that. He'd never play with any other toys."

Winston was sure playing with this one.

"That's not like Winston," Andy commented and shook his head. "He's not making those growling noises. He always growls and hisses when he plays with his toy."

I looked up at the two of them. "He growls and hisses?"

"Oh, yeah," Andy answered. "Every time we play."

As we were talking, Winston and I were engaged in a lively game with the Play-n-Squeak. He stalked, crouched, and pounced on the toy as I moved it along the carpet. For a cat who never played with anything but his "special" toy, he was doing a pretty good job. He was having fun. I was very concerned about what kind of play the McMillans were providing for Winston. Hissing is not a part of playtime.

"Can I see Winston's toy?" I asked casually. I was worried that they might be overstimulating him with whatever they were using. Then, if they attempted to introduce Courtney, he'd be too aroused.

"You go get it," Barbara instructed Andy and then turned back to me. "You see, Winston loved to play as a kitten, but he played too rough. He'd leave scratches and bites all over our hands. So we came up with this idea, and it's worked."

While Andy was in the other room getting Winston's toy, I shifted my position on the floor. As I put the Play-n-Squeak down, Winston trotted over and rubbed up against me. A loud, steady purr came from him. He had obviously enjoyed our game and was thanking me. I reached out and petted him.

"He's usually nervous around strangers," Barbara remarked. "He must like you."

Andy reappeared in the room, holding his right hand behind his back.

"Oh, Winston, look what Daddy has," he announced and knelt down behind the chair. I watched as what looked to be two orange ears appeared around the side of the chair. Next, the top of an orange head became visible. As more of it came into view, I saw what the toy was—some kind of orange cat puppet. Andy was making it dance and dart along the carpet. He was laughing and encouraging Winston to come "play."

"Isn't it a great idea?" giggled Barbara. "Winston attacks it like crazy. With you doing cat behavior and all, maybe you could use this with your clients."

I avoided looking at her because I was afraid my face would give away the horror I was experiencing.

Suddenly I heard Winston hissing and growling. Actually, it was more like a scream. He was instantly on the puppet, biting, scratching, and kicking with his hind legs.

"Andy, please stop," I said.

"What?" He looked up, puzzled.

"Take the puppet away, please. Winston isn't playing. He's being aggressive."

"No he's not," defended Barbara. "We've done this since he was a kitten."

Winston's posture, actions, and vocalizations were all associated with aggression. For two years the McMillans had conditioned him to attack this puppet. I now had a good idea why he was attacking Courtney.

In my work, I try never to place blame on the owners because I know they're doing what they believe is in the cat's best interest. Even when the reason for the problem should be obvious, I know it can often be overlooked by a well-meaning owner. With that in mind, I patiently proceeded to educate Barbara and Andy on how to read their cat's body language. We went over the differences between how he played with the Play-n-Squeak compared to the puppet.

The McMillans were completely surprised to learn that, all this time, Winston's reaction to the puppet had been neither playful nor fun. But as we continued to talk, they began to understand why he wasn't so eager to embrace the new kitten as a friendly companion.

"I feel so awful," Barbara finally said as she looked over at Winston.

"Boy, do I feel stupid," Andy admitted. "Will we have to get rid of Courtney?"

"Let's go over a behavior-modification plan first," I suggested. "Right now, Courtney can stay in the bedroom. We need to work on Winston."

We spent the next hour going through all the interactive toys and how to use them with Winston. Playtime, I explained, needed to be a fun and positive experience for a cat—not, as was the case with the puppet, a need to battle for safety and territory. The cat puppet was an imposing enemy that kept reappearing on Winston's turf. No matter how hard he tried, he couldn't get rid of him. The puppet, which was quite large, would suddenly appear, startling Winston. Andy admitted that he would often pin the cat down with the puppet, forcing him on his back. He thought Winston was enjoying the game.

As a child, Andy had had a dog who loved to play rough with him. Never having had a cat before, he played with Winston the same way. Barbara had only had outdoor cats in the past, and she admitted she never had played with them. Putting out food and water had been the extent of her previous feline interaction.

At one point, Winston walked over to one of the toys by the chair and pawed at it gently. I took the hint, picked the toy up, and we resumed our game.

During our session, Barbara and Andy learned how to use playtime as a chance for Winston to enjoy his predatory skills. The enticing little toys dangling from each wand represented prey. The key was to make the toy act the way prey would.

"This toy," I began, holding up the little dangling object, "can represent a mouse, cricket, lizard, or anything a cat might hunt. Compare the size of this to the intimidating size of that puppet cat. This little toy here will bring out the hunter in Winston, not the fear."

The McMillans each took a turn using one of the toys with their cat. He responded with beautifully timed pounces and successful captures.

"To build Winston's confidence and keep playtime enjoyable, be sure he makes plenty of captures during each session," I said, watching Winston as he proudly held the toy in his mouth.

After our play-therapy session, Winston went over to one of the chairs, jumped up, and stretched out. He then began a leisurely face-washing. Contented, he fell asleep.

It was time for me to meet Courtney.

Andy opened the door and motioned for me to enter the bedroom in order to see the source of Winston's woes. As I stepped into the room, it became immediately clear why poor Courtney had never stood a chance with Winston. There, curled up on the bed, was a living miniature version of the puppet. Courtney looked exactly like the very thing the McMillans had taught Winston to hate.

As soon as Courtney heard us in the room, she sat up and peered behind us, as if to check on Winston's whereabouts. When she was certain he was not around, she relaxed and jumped off the bed to meet us.

After spending time with Courtney, I provided the McMillans with a full-scale plan for reintroduction of the two cats. For the time being, they were to continue the room/cat switching. Also, twice daily they were to have gentle play-therapy sessions with Winston. I stressed the importance of this. I wanted him relaxed, happy, confident—and never overly stimulated.

I discussed the possibility of using a mild anti-anxiety medication on Winston for a brief time, if necessary, during the reintroduction phase. "We're going to try it without the medication first," I said. "But if one is required, we'll discuss it with Winston's veterinarian."

Next, Andy was to install three baby gates (on loan from me) in the doorway to Courtney's room during the time the McMillans were home. The gates were to be installed one above the other, to prevent Winston from getting in and Courtney from getting out. This would also accustom Winston to seeing Courtney on a regular basis, but always in the same location. There would be no surprise appearances like there had been with the puppet. Winston would know that he was safe and in control. We had to break the behavior pattern.

It was also important to surround Winston with positive distractions so as not to frustrate him, since he couldn't get to Courtney. During our session, I asked Barbara if they ever gave any special treat food to Winston.

"He loves chicken," she said. "We treat him to some now and then."

Now, I normally don't believe in feeding table food to cats, but during difficult behavior sessions, a little bit of treat food (and I do mean just a little) can help cats associate good things with each other. So if Winston got a few nibbles of chicken in the hallway outside of Courtney's room, it would contribute to letting him know that *good* things were happening when the little kitten was around. That, combined with play sessions and lots of attention, would enable him to feel less threatened by the presence of the newcomer.

Armed with a few more specific instructions from me, the McMillans felt a little more optimistic about Courtney's future in their home. As I left, I assured them that I was just a phone call away. I wanted regular progress reports.

After five days of regular play-therapy sessions with Winston, using the proper interactive toys, we installed the baby gates to Courtney's room. Andy reported that Winston hissed and growled. Hissing in that situation was to be expected and quite normal, but with the help of the chicken treats and gentle playtime, Winston soon quieted down. After the gates had been up for four days, the hissing became limited to just a few sporadic episodes. By the fifth day, Winston would walk by the room and stop, but without so much as a hiss.

One week later I received a follow-up call from Barbara to report that Winston and Courtney had touched noses through the gates. Winston had even gingerly put his paw through the gates (politely, without claws) and touched Courtney.

"I'm so excited," Barbara was almost crying. "I never thought this would happen. I was sure we'd have to give Courtney up."

Just days later, Courtney and Winston were regularly engaging in play through the gates.

"It's time to open the gate and let Courtney out," I said during the next follow-up call.

"I'm too nervous," Barbara admitted.

"Just be casual about it and Winston won't view it as a big deal."

Barbara and Andy were very worried that Winston would attack Courtney if given the chance, so I agreed to come over.

The plan was for me to arrive at noon, stay for lunch, and conduct a play session with Winston. That way he'd be used to having me around before we opened the gate.

The big moment was upon us. Barbara had a small container of cooked chicken pieces. "In case of an emergency," she said.

"It'll be fine." I patted her on the arm and told both of them to go into the living room; Winston didn't need the two of them staring at him, just waiting for him to make a wrong move.

I sat down in the hallway and casually clicked open the bottom gate. Courtney cautiously stepped out as Winston looked on. He then stood up suddenly. I watched out the corner of my eye, trying to appear very casual, as Winston walked behind Courtney and sniffed the opened gate. Naturally, he noticed the plastic container of chicken sitting close by.

"Well, I'd say this exceptional behavior deserves a treat," I announced, and offered him a piece. I then reached over and gave a tiny bit to Courtney.

Courtney walked to the end of the hallway, near the living room entrance, and waited. It was as if she knew she needed Winston to grant permission for her to proceed into his territory. Winston approached her. I held my breath, not moving a muscle. He stood nose to nose with her, then put out his tongue and licked her along the top of the head. Satisfied, he walked out of the hallway. Courtney looked at me and then over to Winston's departing figure.

"Well, go on, Courtney," I smiled. "I think he just gave you his seal of approval."

○●○

Eggplants Don't Lie

Joyce Knight and Gwen Carson were next-door neighbors in a quiet little subdivision just outside of Nashville. Joyce and Gwen had been living next door to each other for six months. Gwen, a divorced mother of a fourteen-year-old daughter, had a cat. Joyce, who was also divorced and the mother of a thirteen-year-old girl, was a dog owner. Recently, the two neighbors had been feuding.

It all started when Joyce opened her back door one morning shortly after Gwen had moved next door, and noticed one of her hanging bird feeders had been knocked to the ground. Thinking the previous night's wind may have blown it from where it had been hooked to a branch, she at first thought nothing of it. But the following morning when she discovered her birdbath knocked over and some feathers lay scattered on the ground, she suspected more than the wind. She began suspecting Gwen's cat, who had developed a morning habit of sitting on the fence between the two yards. Previously, Joyce hadn't had any trouble with cats in her yard, despite her numerous bird feeders.

On the other side of the fence, Gwen was beginning to have her own suspicions as well. It seems that someone had trampled in her garden, knocked down the small chicken-wire fence, and uprooted many of her vegetables. Ironically, her suspicions centered on Joyce's dog. Gwen had noticed that the dog was often standing on his hind legs, straining to see over the wooden fence. On several occasions, she'd even found him lounging on her patio, basking in the sun.

Snowball, Gwen's white cat, and Mosely, Joyce's black Labrador retriever, became instant friends. Snowball would sit on the top of the fence that separated the two yards and Mosely would stand up on his hind legs so he could reach her. She'd rub against his face and lick his nose. Mosely was patient and extremely gentle with his little friend. Gwen and Joyce, who were becoming friends themselves, originally didn't mind the visits from each other's pets because they both seemed well mannered and friendly. In fact, everyone appeared to be pairing off quite nicely. Joyce and Gwen both enjoyed many of the same interests and their children had immediately taken to each other as well. Within two weeks, Gwen's daughter, Megan, and Joyce's daughter, Susan, were inseparable.

Life seemed idyllic for both families. Joyce was so grateful that Gwen had moved into the neighborhood. Ever since her husband had left, she and Susan had been lonely. The family who had previously lived next door didn't have any children and hadn't been sociable to Joyce.

For Gwen, moving to a new home after going through a bitter divorce was frightening. Gwen was so worried about taking Megan out of the school she knew, but they could no longer afford the home they'd lived in when Charles was around. So Gwen was relieved to find such a perfect neighbor in Joyce and her daughter. Things were going to be okay.

Not quite. It was two months later that Joyce began noticing the disruption to her bird feeders and Gwen was finding her garden uprooted.

Initially, Gwen just quietly replanted her garden and tried to reinforce the surrounding chicken wire. After all, she wasn't positive that Mosely had been the culprit. Unfortunately, though, the evidence quickly mounted against him the next morning when Gwen found him standing in the middle of the once-again demolished eggplants and zucchini. New friend or not, Gwen was going to have to tell Joyce to do a better job of confining her dog.

With what was left of a trampled young eggplant in her hand, Gwen walked back into her house, told Megan she'd be right back, and started for her neighbor's house. She only got as far as the front door when she saw Joyce standing on her porch, just about to ring the bell. The expression on Joyce's face was very serious. *Well, Joyce,* Gwen thought, *your mood is about to get even worse.* Gwen opened her door, the eggplant in her one hand. "Joyce, I was just on my way over to your house."

"Gwen," Joyce said, her expression very stern, "we have a problem."

"We sure do." Gwen nodded and held out her hand to display the eggplant. "Your dog is the problem. He destroyed my garden."

Joyce's eyes widened. "My dog's the problem? I don't think so!" she said and held out her own hand, which contained a dead bird. "Your cat is killing my birds. *She's* the problem, not my dog."

The two neighbors stood face-to-face, with their hands outstretched, one with a rotting eggplant, the other with a stiff bird.

"Snowball never hunts birds," Gwen said, breaking the tense silence.

Joyce raised the bird up closer into Gwen's view. "Oh, really? Well there's a first time for everything."

"What about my garden?" Gwen shot back.

"Well, what about my birds?" replied Joyce. "Why can't you keep your cat indoors?"

Gwen shook her head. "I can't. She loves being outdoors. Besides, she doesn't hunt birds!"

"Well, she's hunting them now. Just try keeping her indoors for a while."

Gwen looked at the little dead bird and felt sad. Could her sweet, docile cat, who had never shown any interest in hunting, suddenly have turned into the bird terminator? "I'll try," she said and immediately saw Joyce's face soften a bit. "But will you keep your dog in his own yard?"

Joyce realized that she'd have to make some concessions if she wanted to save her birds. "Yes," she agreed.

The two women went about their day, each feeling relieved that the problem had been solved. In reality, though, the problem was about to go from bad to worse.

For the next week, Snowball was confined indoors—something she was not at all happy about. She sat by the window for hours each day, looking longingly toward Mosely's yard. Often, a loud, mournful yowl could be heard whenever she spotted Mosely being let out the back door. After taking care of his personal duties, the dog would run to the fence, balance on his hind legs, and look over the fence for his feline friend. When he'd spot her in the window, his tail would wag wildly and he'd bark excitedly. Joyce, who diligently stood watch on the back porch, would call him back into the house. Reluctantly, Mosely obeyed. Once inside, he'd go to the window and sit, gazing across the yard at Snowball's house.

Joyce and Gwen thought they had their gardening and wildlife problems solved, although their own pets seemed less than thrilled with the new arrangement. The two women decided that it would take time, but Mosely and Snowball would eventually adjust to their restrictions.

Then, one evening, as Joyce was opening the back door to let Mosely out for his late-night walk, she noticed something out of the corner of her eye when she switched on the light. On the bottom porch step was a brightly colored object. She walked over for a closer look and found a pile of red feathers. "Snowball!" she said disgustedly as her eyes went from the feathered remains toward her neighbor's house. "I'll deal with you tomorrow," she said in the direction of Gwen's home. The next morning, before going off to work, she'd just pay a little visit to her *former* friend and complain about the fact that she was obviously not adhering to the agreement.

In the morning, armed with a dustpan of red feathers, Joyce marched over to Gwen's house and banged on the door. Gwen's daughter answered the door.

"Hello, Megan, where's your mother?"

"Uh, in the backyard," answered the girl, staring wide-eyed at the contents of the dustpan. She then held the door open as Joyce marched through the house, headed for Gwen's backyard.

Joyce found Gwen kneeling in her garden . . . or, rather, what used to be a garden. Something or someone had trampled the beds again.

"Good morning, Gwen," Joyce said coldly.

Gwen looked over her shoulder, then stood up, brushing the soil off her clothes. "You broke your promise. You let that dog ruin my garden again," she said angrily, picking up a fistful of broken stems and squashed vegetables.

"What are you talking about?" Joyce asked impatiently.

"This," Gwen answered, shaking the eggplant, causing bits of it to come loose and hit herself in the face.

"You're a fine one to talk," Joyce responded and held the dustpan out in front of her.

"Yuck, what the heck is that?" Gwen asked, taking a few steps back.

"It *was* a bird," Joyce responded, "now it's a corpse."

For the second time, the two neighbors stood face-to-face, one holding a very dead bird and the other holding a very dead eggplant.

The argument grew heated as each swore to the other that they hadn't broken the agreement. Gwen didn't believe Joyce. And Joyce, staring at the red-feathered evidence belly-up in her dustpan, certainly didn't believe Gwen, either.

"Mom?" said a little voice, struggling to be heard above the rising voices of the feuding neighbors.

The two women momentarily stopped their argument and saw Gwen's daughter standing in the doorway. Joyce's daughter was right behind her.

"What are you doing over here?" Joyce asked her daughter.

"It's not Mosely or Snowball," Susan called out.

"How do you know?" asked Gwen.

"We've seen a huge cat hanging around," Megan said, pointing toward Joyce's backyard. "He chases the birds in your yard, Mrs. Knight."

"It's true, Mom," Susan added sheepishly.

Joyce started walking toward Gwen's back door, still holding the dustpan. "How come you didn't tell me? When did you first see him?"

"Yesterday," replied both girls in unison.

Joyce looked down at her dustpan and then back up at Gwen. She smiled. "I'm so sorry."

Gwen tossed the dead eggplant back into the garden and wiped her hands on her pants. "I am, too."

Both women felt greatly relieved, but they still had a problem on their hands. How were they going to protect Gwen's garden and Joyce's birds?

"I have an idea," Gwen said later that morning as they sat having coffee in her kitchen. "This'll be perfect." She got up and looked through her purse. "Where'd I put that . . . oh, here it is," she said as she pulled out a slip of paper containing a name and phone number—mine. "The vet gave me the name of a, get this, *feline behaviorist*, because Snowball had been so depressed now that she can't go outside. I was going to call this woman to see how much she charged."

"A cat shrink?" Joyce laughed.

"Well, I was desperate," Gwen responded, "but lately, Snowball's been starting to act normal again, so I wasn't sure whether to go ahead with it. I might as well do it now because she can maybe help us trap that stray cat."

Joyce nodded in agreement but was still chuckling over the thought of a cat shrink, even while Gwen dialed the phone.

Four days later I was standing in Gwen's backyard, listening to both she and Joyce explain the sequence of events. Since the time Gwen had originally called me, there had been another attack on Joyce's birds.

Although I don't do any trapping and rescue work myself (I refer clients to a network of several well-qualified people), I agreed to offer any suggestions I might have, since I was going there for a consultation on Gwen's cat.

It certainly was a frustrating situation for both women. Standing in Gwen's garden, which by now was just a sea of stems, I could understand how upset she was. It was obvious that she had put a lot of love and care into her garden. I looked over the trampled stems, uprooted plants, and half-eaten vegetables. Two things confused me, though. One was the condition of the chicken-wire fence. A cat would've just climbed over it, not torn the fence down. The munched-on vegetables also puzzled me. I turned to Gwen and Joyce. "You're sure that your daughters said they actually saw a cat in this garden?"

"Yes," Gwen answered. Joyce nodded in agreement.

"Was the cat digging in the soil or actually eating the vegetables?" I asked.

"I'm not sure. I'll go get the girls so they can tell you themselves," Gwen said, and then started back toward the house. As she was leaving I noticed that two girls in an upstairs window were watching us. The expressions on their faces appeared quite serious.

Joyce turned my attention back to the immediate situation. "I live right there," she said, pointing to the house on the other side of the wooden fence.

I walked over to the fence and peered into her yard. There were about a dozen bird feeders and several birdbaths. Now this was a backyard that would appeal to a cat—not Gwen's vegetable garden.

I heard a voice behind me and turned to see that Gwen was back with the two girls. After the initial introduction, the girls immediately fell silent.

"Tell Mrs. Johnson-Bennett what you saw," Gwen prodded Megan.

"We saw a big cat," Megan responded flatly.

"Did you see the cat in the garden?" I asked.

Megan nodded.

"What was the cat doing?" I asked.

Megan looked at Susan, who then answered, "Eating."

"Eating the vegetables?"

"Yeah," was the reply.

I walked over to the garden, knelt down, and examined the eaten vegetables. "And you're sure it was a cat and not a rabbit or dog? Perhaps even a deer?"

"It was a *cat.* A big orange cat," Susan answered defensively while appearing to avoid looking at me.

Something strange was going on. The two girls were becoming very fidgety. I know that children can certainly be fidgety just for the sake of being fidgety, but this was something else. I wasn't sure what was going on, but I was looking at a demolished garden—an act that seemed highly unlikely to have been accomplished by a cat. I doubted that a cat would use this garden for anything other than a giant litter box. It was very doubtful that being the true carnivore that a cat is, he would opt to eat zucchini, tomatoes, and eggplant. And even though a cat *may* develop a preference for these veggies, the size of the bite marks pointed to a larger animal.

The girls may very well have seen a cat in the garden. They may have even seen him heating vegetables, but something else had made these large teeth marks.

I wasn't getting the answers I needed, so I decided just to instruct Gwen and Joyce to call a local rescue group to help them trap the intruder—whatever it was. A cat may have been hunting Joyce's birds, but no cat, no matter how big, was helping himself to Gwen's vegetables.

"What could it be?" Gwen asked.

"I think it's a dog," I answered.

Gwen shot a harsh look at Joyce.

"Don't look at me," Joyce defended, "Susan and I watch Mosely every time he goes out. Isn't that right?" she added, looking over at her daughter, who silently nodded.

"Well, it's getting chilly out," Gwen said. "Let's go in the house and finish this discussion."

As we walked toward the house, I noticed that Megan and Susan kept whispering to each other and then looking at me. Considering the fact that these two were meeting a "cat shrink" for the first time and that they'd just seen me closely scrutinizing rotting vegetables, I figured some whispering was probably called for. But my intuition kept telling me there was more here than meets the eye. The amateur detective in me started perking up.

Over tea, Gwen and I got down to business about Snowball's on-again off-again depression. Joyce asked to sit in (just for laughs, I suppose), but the two girls bolted out of sight the minute we came inside.

Gwen filled in the details about how sad Snowball had become once she was confined indoors. "It seems to have improved, but I'm still concerned," she said. "I don't want her to think she's in jail."

"How long has she been indoors now?" I asked.

"Almost two weeks," Gwen answered.

"Can I see her?"

"Sure," she said. "She's probably upstairs with Megan."

Joyce waited in the kitchen while Gwen and I went upstairs to Megan's room.

Megan's door was closed. Gwen knocked and then entered. Both girls were sitting on the floor, huddled together in deep conversation. Snowball was curled up on Megan's bed.

"Mrs. Johnson-Bennett would like to see Snowball," Gwen said to the girls.

I smiled at both girls but no one smiled back. Instead, I received two icy cold stares. *What the heck were these girls up to?* I wondered as I quietly walked toward the bed to see the cat.

Snowball awoke, stood up, stretched, and then walked right over to the edge of the bed, soliciting my attention. I let her sniff my hand and then reached out to pet her. She rubbed up against me, purring. I petted her again but as I ran my hand across her fur, I came upon something pinchy. I parted her medium-length fur and found a couple of pieces of twigs caught up in her coat.

"When did you say was the last time Snowball had been outside?" I asked Gwen again as I removed the twig pieces.

"Two weeks. Why?"

I was about to answer her when I spotted something else. Inbetween Snowball's left front paw was the tiniest piece of a blue feather. It was caught in her nail. "Joyce told me outside that she'd found another dead bird recently. Did she say what kind?" I asked Gwen casually.

"A blue jay, I think."

I noticed that Megan and Susan had exchanged glances and then got up and left the room.

"Gwen, I'd like to spend a minute alone with Snowball. Would that be okay?"

"Sure," she agreed, somewhat puzzled. "I'll be down in the kitchen with Joyce," she added and then left.

I sat on the edge of the bed, petting Snowball and checking over her fur. After a minute or two, I had the feeling that we weren't alone. I looked up and saw Susan and Megan watching me from the doorway. I figured they'd be back. It was time to solve this puzzle. "Come on in, girls," I said quietly. "We need to talk."

Reluctantly, Megan walked into the room and Susan followed.

"Snowball and Mosely are pretty close, right?" I began.

Both girls nodded.

"Did it make you sad when they weren't able to see each other anymore?"

"Sure," Susan answered. "They love each other."

"I know how hard it must've been for you to watch your pets become so sad and lonely. I love my pets very much and it breaks my heart whenever they're sad. I'd do anything to make them happy," I said, and then held out the piece of blue feather in one hand for the girls to see. "I found this on Snowball. I think it's a blue jay feather. What do you think?" I paused as both girls looked at the feather and then back at me. "You know how it got there, don't you?"

Silence.

"Your mothers are very upset over this situation. You both are as well. And of course, Mosely and Snowball are extremely sad. I can help all of you, but I have to know the truth," I gently said.

Megan looked at Susan and then at me. "We've been—"

"Quiet!" Susan interrupted, giving her friend a nudge.

"Do you really want to help Snowball and Mosely?" I asked.

"Of course," Megan replied.

"Then let's tackle this problem together—the right way. Are you afraid of what your mothers will say?"

Megan weakly answered, "Yeah."

"Okay, then let me help you. Tell me the truth."

Over the next few minutes, Megan and Susan told me how they couldn't stand to see their pets so miserable. The animals hated being apart, and Snowball resented not being able to go outside. So the girls had started letting the pets outside as soon as their mothers left for work, and then bringing them in before they returned home. Unfortunately, Mosely began destroying the garden again and Snowball killed more birds. The girls had made up the story about the stray cat to get their mothers to stop fighting.

"They were yelling at each other, so we had to do something," Susan admitted.

"You both realize that you have to tell your mothers the truth," I said. Both girls looked at me in horror, but I continued on. "Let's do it together. That way, I can help with a solution to this problem."

Megan and Susan finally agreed and the three of us marched downstairs to face their mothers.

After the confession, which went rather well, considering how Joyce felt about her birds and Gwen about her garden, punishment was discussed. Gwen and Joyce opted for grounding. I suggested that instead of grounding, Megan and Susan be given specific responsibilities to help each pet through this process.

First of all, I explained that Snowball had to remain an indoor cat because the temptation of Joyce's birds was far too great. But in order for it not to be a kitty prison, Gwen and Megan had to engage the cat in regularly scheduled interactive play sessions. This could be part of Megan's responsibility. Whenever you restrict a cat indoors, you have to make that new life as exciting as the previous outdoor existence had been. I explained to Gwen and Megan that Snowball was bored and they needed to spice up her life. My other suggestion included bringing Mosely over for regular visits. I believed that Snowball might also benefit from the addition of a feline playmate. This idea thrilled Megan, and she began squealing at the thought. Gwen, surprisingly, took to it as well.

For Mosely, I suggested regular playtime and exercise (this would be Susan's job). I also felt that Mosely would benefit from the addition of a companion. A second dog would eliminate Mosely's loneliness. "If you get him a companion dog, they'll be able to romp outdoors together and I bet he'll probably stop trying to jump the fence to Gwen's yard," I offered. I felt that Mosely was raiding Gwen's garden due to boredom.

"We don't play with him much," Joyce admitted.

I turned to Gwen. "And Snowball's depression is due to the loneliness and boredom of being confined and denied access to all the fun of hunting birds and playing with her friend. You said that her depression had improved and that was because Susan and Megan were letting her outdoors again," I said. "Something tells me that Snowball doesn't get much playtime or exercise in the house, does she?"

"No, she doesn't," Gwen said.

To protect Gwen's garden from any other possible unwanted visitors, and also to eliminate future temptation by Mosely, I recommended that she invest in a sturdier enclosure—perhaps even have one built. She said she'd call the local home improvement store in the morning.

"So, do we have a plan?" I asked.

Everyone said yes.

With the mystery of the phantom eggplant-eating cat solved, Gwen could have a garden again, Joyce's birds would be safe, Megan and Susan could stop conspiring, and Mosely and Snowball would no longer have to gaze longingly at each other through the windows.

Not too long after my visit to Gwen's house, I received a gift basket from her that contained beautiful home-grown vegetables with not a toothmark on them.

○●○

The Divorce

"This is Allen Hetherington," the message said. "I need to speak with you concerning a legal matter. If you would, please call me tomorrow morning before I have to leave for court."

A legal matter? I jotted down his name and number. He wasn't at all familiar to me. I was curious—and slightly concerned.

The next morning, I returned Mr. Hetherington's call. His secretary put me right through.

"Mrs. Johnson-Bennett, thanks for calling back," he said.

"What can I do for you?"

"A client I'm representing in a divorce has suggested that you might be of assistance concerning the couple's cat."

"Does the cat have a behavior problem?" I asked.

"No," he answered. "The problem is with the couple. My client and his wife have agreed on all aspects of property division, except where it concerns their pet. They found this cat during their marriage, and clearly it belongs to both of them. The problem is they can't come to an agreement on which party should keep it." His tone of voice indicated that he found this all a complete waste of his time.

"What is it you need from me?" I asked.

He cleared his throat. I heard him shuffling papers in the background. "Both my client and his wife have agreed that they'd appreciate your input."

"My input?"

"Yes." The sound of rustling papers stopped. "We'd like you to spend some time with the couple and their cat and offer an opinion as to where you think the cat would be happiest."

I'd never been asked to do anything like this before. "Mr. Hetherington, I'm not sure I should . . ."

He interrupted, "Please, Mrs. Johnson-Bennett. This couple has been completely unable to resolve this one issue. They're both willing to abide by whatever you say." He sounded impatient with me.

I heard a female voice in the background and the sound of him cupping the phone with his hand. Their conversation was muffled, then he was back on the line. "Excuse me, Mrs. Johnson-Bennett. I must leave now for court. Would you be willing to meet with the couple?"

"Yes, all right."

"Thank you. That's wonderful. I'll have my secretary call you to set up the meeting." And then the line went dead.

Mr. Hetherington's secretary rang me twenty minutes later. "I was wondering," she asked, "where you would prefer the meeting to take place—the client's home, here at the office, or at your office?"

"I'd prefer the cat to be in a familiar environment. The client's home would be my recommendation."

"Very good," she said, sounding relieved. I suppose her boss wasn't anxious to have a cat brought in to his office for a meeting.

The appointment was scheduled for the following Tuesday. Both husband and wife would be at the house. (The husband had moved out several months before). No lawyers would be present (at my request), so the cat wouldn't be disturbed by too many unfamiliar faces. Additionally, the couple agreed to remain civil to each other in order to keep our meeting productive.

Debbie and Mark Kelbreth were divorcing after eight years of marriage. Since they had no children and were having little trouble splitting their accumulated possessions, the divorce had thus far gone smoothly. Mark had agreed that Debbie could have the house, and she would then take over the mortgage payments. They each earned a good living, so there was very little argument over financial arrangements. The only thing they were still arguing about was who would get custody of their five-year-old cat, Misty. Both Debbie and Mark loved her dearly, and neither was willing to give her up. Since the separation, they'd been temporarily sharing custody of her.

Debbie's lawyer had recommended joint custody on a permanent basis, but Mark was planning to move back to his hometown, which was more than one hundred miles away. There seemed to be no compromise that would satisfy both parties. They each wanted the cat, and they were hoping for me to provide some insight. Believe me, I felt the pressure.

The background I received was that Misty, a purebred Manx, had been purchased from a breeder, although Debbie hadn't wanted to get a purebred cat. She'd hoped to fall in love with a homeless stray from the local shelter, but Mark was adamant about having a purebred. So they visited several breeders until they found the right kitten.

Shortly after bringing their new kitten home, Misty became seriously ill. The vet recommended that they return her to the breeder, but the couple was already strongly attached to her. Debbie was determined to do whatever it took to make Misty well.

After weeks of round-the-clock care, Misty finally began to grow stronger and healthier. Both Mark and Debbie had devoted all their time to saving the new member of their family. Now, five years later, neither one was willing to say good-bye.

Our session was conducted on their sun porch. It was a beautiful setting and, under different circumstances, would have made for a lovely visit. I could see how nervous Mark and Debbie were. I was nervous myself, since this was my first custody case.

I became very sad as we talked. It was obvious that both Mark and Debbie adored Misty and that this was a terribly painful situation. Considering that this young couple were about to dissolve an eight-year marriage, they were remarkably caring toward each other as they explained why they were separating—something I didn't need to know, but they felt compelled to offer. It was only when Misty's name came up that the tension between them was obvious.

As we chatted, I noticed Misty sitting at the far corner of the porch looking out through the screen at the birds in the backyard. She had no idea that she was the center of so much turmoil.

I spoke with the Kelbreths about their five years with Misty. Owners are usually very willing to share cat stories with me, so I was hoping that by letting them talk I'd get a clearer picture of who these people were and what kind of relationship they each had with Misty.

The discussion got competitive when I asked who had handled specific responsibilities for Misty while the couple lived together. They started arguing over who fed her more, who took her to the vet, who sat up with her when she was sick, and who brushed her most often. I looked over at Misty, whose ears were twitching back and forth. Poor cat, her name was being mentioned so much, she probably didn't know what to make of it all.

To ease things, I suggested we spend some time with Misty. Debbie brought out Misty's toys, and we all sat on the porch floor as they each tried to entice the cat into play. Misty wasn't at all interested, and I didn't blame her. She was refusing to be a party to this game of one-upmanship. I took over. I showed the Kelbreths a few of the interactive toys I use in my work and engaged Misty in a leisurely game.

At the end of the session I asked if I could see them each again, only this time one-on-one. I needed to get beyond the arguing over

Misty so that I could get a clear picture of how each interacted with the cat. Both of them looked extremely worried.

We set up an appointment for each person. Debbie's was scheduled for the next day and Mark's was two days later. It was agreed that Mark would come to the house for his appointment so Misty wouldn't have to be shuttled back and forth. Debbie said she'd go out for the afternoon.

The following morning I went to Debbie's house for our session. She eagerly greeted me at the door. "I'm very optimistic," she said brightly as we walked back to the porch. "After all, Misty will do much better staying in the home she knows, rather than moving in with Mark." She was upbeat and confident, not the way I remembered her from the previous day.

Initially, my plan was to spend time talking with Debbie and then let her engage in a play session with Misty. But the odd thing was, from the time I walked in the door, Misty was literally all over Debbie. She rubbed against her legs as she walked, and when we sat, she sprang up into Debbie's lap, rubbing her chin vigorously along Debbie's arm. This definitely was a departure from the way the cat had acted the day before. Hmm, something wasn't right. I wondered if there was some manipulation going on here.

"Is Misty normally so affectionate with you?" I asked casually. "Because she didn't seem this way yesterday."

"Oh, yes," answered Debbie. "It's just that she's more reserved when Mark's around. That's why I'm glad you wanted to see us alone."

Funny—when I'd first mentioned doing a one-on-one session, both Debbie and Mark had appeared very apprehensive. I smelled a trick.

As we talked, I watched Misty rolling and rubbing against her owner, and my suspicions continued to rise. "Debbie," I began gently, "I know Misty means a lot to you and that this is a very difficult situation. But the way she's behaving is very similar to the way a cat responds to catnip. Did you give her some before I came?"

She didn't answer, just looked innocently at me. Oh, yeah, I was being tricked.

After a little more prodding, Debbie finally admitted she'd rubbed catnip along her arms and on her clothes when she saw me pull in the driveway. "I had a feeling you'd figure it out," she said.

With that out of the way, we got down to business. We talked about the special qualities in her relationship with Misty. She told me how she'd nursed her back to health when she was a kitten. Debbie also shared with me many stories of life with Misty and how much she'd miss her if Mark took her away.

Our visit ended after an hour, and I returned two days later for my appointment with Mark. Now, this was strange—but, from the moment Mark answered the door, there was Misty, following him everywhere, just as she'd done with Debbie. She jumped up in his lap when we sat down and started licking his arms and face. I couldn't believe it, but I was seeing a rerun of the previous afternoon. However, there was also a slight difference in the way Misty was responding. This definitely wasn't a catnip reaction, but *something* was going on nonetheless.

As Mark chatted away about his feelings for Misty, I tried to focus on what the cat might be reacting to. It had to be some kind of food. I leaned forward slightly as Mark talked, trying to get a whiff of anything that could give me a clue. But as I leaned forward, Mark leaned back. I leaned forward just a bit more and, sure enough, Mark leaned back even farther. He was trying to avoid detection. I sniffed the air. I couldn't smell anything. I sniffed again. My first thought had been that he might have used tuna—something most cats go wild for—but I couldn't detect any fishy smell. I leaned forward one more time. Mark was already leaning as far back as he could. He was trying very hard to appear nonchalant.

"Misty seems extraordinarily affectionate toward you. Is she normally this way?" I asked with a raised eyebrow.

"With me she is," he answered. "She's much closer to me than to Debbie."

Okay, I'd had enough of this. I was tired of playing games. I got a little tough with Mark and told him he'd better come clean about whatever he'd done to influence Misty's demonstrative behavior. With a little badgering, Mark eventually confessed that he'd rubbed his arms and face with some cooked chicken he'd brought over in a plastic bag from his apartment.

He and his estranged wife sure had a lot in common. They thought alike. Too bad they weren't staying together.

Once the truth was finally out, we continued our session.

I had spent quite a bit of time with both Mark and Debbie, and now I had to carefully consider which one would be best for Misty. It was an extremely difficult decision. Even though they'd each tried to trick me, I knew it was out of their desperation to keep the cat. I struggled, but was finally ready to offer my opinion.

It had been two days since my session with Mark. I had planned to call both Mark's and Debbie's attorneys after my morning appointments. I came home for a quick lunch and was checking my phone messages, with the intention of placing a call to Mr. Hetherington as soon as I finished. But on my machine was a message from him. "He's probably anxious to get this resolved," I said out loud as I dialed his number.

"Mrs. Johnson-Bennett, thank you for returning my call so quickly," the lawyer said once his secretary had put me through.

"Mr. Hetherington, I'm ready to . . ."

He jumped in. "Yes, yes. That's the reason I called you. It seems that Mr. and Mrs. Kelbreth have decided to make an attempt at reconciliation."

"Excuse me?" I put down my sandwich.

"The events of the past few days apparently have brought up some emotions they've been unable to ignore." He exhaled deeply. "I apologize for any inconvenience on your part. Please send me a bill for your time and I'll forward it on to the Kelbreths." He sounded relieved to have this case behind him for the time being.

After I hung up the phone, I sat there for a long time. It was very strange. Did the events of the past few days involving Misty truly influence how Mark and Debbie felt about each other?

In my career I've helped repair relationships between cats and owners, and I know I've helped ease tension among family members during a feline behavior crisis. But as much as I'd love to take the credit for saving this couple who were on the brink of divorce, I honestly have to give credit where credit is due—to Misty.

Oh, by the way, if you're wondering what my opinion was regarding who should have kept Misty, I think it's best to keep that answer locked away in my files for now. Mark, Debbie, and Misty are happy together, and that's all that matters.

Since working with the Kelbreths on my first cat-custody case, I've been asked to arbitrate several others. Unfortunately, most cases don't have this kind of happy ending. All too often, the cats suffer in silence while the husband and wife do battle. Divorce hurts everybody—even pets.

○●○

Bonsai and the Boyfriend

Jill Dunn, a twenty-five-year-old legal secretary, had been dating the man of her dreams for six months. She and her love were blissfully happy. Even her mother and father adored Ed, a twenty-nine-year-old music teacher. Jill's older brother, Tommy, also thought Ed was a terrific guy. There was just one member of Jill's family who didn't approve of the relationship. It was Jill's *other* significant other: you guessed it, the cat!

During our initial phone call, Jill explained that she had lived alone with Bonsai since he was a kitten. From the moment she'd brought her new boyfriend home, the two-year-old chocolate-point Siamese had been very hostile toward him.

"Bonsai means everything to me," Jill said. "And he's never done this with anyone else who comes over."

"It's not that unusual for a cat to be very wary of somebody who suddenly takes up a lot of your time," I told her. "He doesn't understand the drastic change in his life."

The cat had apparently tolerated Jill's previous boyfriends because they never stayed around long enough to pose any problem. But this was Jill's first serious boyfriend, and Bonsai wasn't laying out the welcome mat. Ed just wouldn't go away, and Bonsai was worried.

"I don't know how much more Ed can take," Jill sighed heavily.

I asked her to describe Bonsai's behavior. She said that he'd basically done a lot of hissing in Ed's direction. So far, that was relatively normal for a cat who is anxious about an intruder in the home.

"Has he ever actually attacked your boyfriend?" I asked. No answer. "Jill?" Silence. "Hello, Jill?" Again, silence.

She finally spoke. "Ah, can we talk about it when you get here?"

Uh-oh! "If you'd prefer," I agreed.

As I hung up the phone, I wondered what Jill was so afraid to tell me. Whatever it was, I knew it wasn't good.

Jill and her cat lived in a small one-bedroom apartment. It was the only home Bonsai had known since Jill first purchased him from a pet store at the age of eight weeks. Bonsai was right beside Jill when she answered the door. In typical Siamese fashion, he extended several vocal greetings.

What struck me immediately about the tiny apartment was how jammed it was with furniture. Jill directed me to the sofa, but in order to get there we had to squeeze sideways between two closely spaced tables.

"I apologize for the way the place looks, but Ed just moved in two weeks ago," she explained as I fumbled to get my briefcase and other paraphernalia through the tight quarters without knocking over anything on the tables. "The lease on his apartment expired, so he's staying here until we can find a bigger place. My lease is up in three months. It's a little inconvenient, but we think it's worth it."

They were both going to live in this place for three months? A little inconvenient? There was barely enough floor space to put a foot down.

Bonsai jumped up on one of the tables and looked around. He hadn't been consulted about this change, and it was obvious he didn't like it. He seemed to be searching for his familiar furniture. Gingerly he walked along the table, hopped to a chair, and then crossed to a wicker rocker shoved in a corner. He settled down there and curled up. I figured the chair must belong to Jill.

I took out my notebook. "Tell me about Bonsai."

Jill shifted her position. "Well, he *hates* Ed," she began, picking at an imaginary speck on the sofa. "At first Bonsai just avoided him, but then he started hissing at Ed whenever he saw him. Now it's gotten to where he hates the very sight of him." Jill looked over her shoulder at the cat and then back at me. She was awfully nervous. Clients are often tense at first, I assume because they fear I'm going to tell them they're bad parents. But Jill's nervousness seemed different.

"Go on," I said, very curious.

She continued. "When we started dating, Ed never stayed overnight here because he's allergic to cats. It worked out great at first, because he had an apartment just a few blocks away. But then his lease expired, and he didn't want to renew it since we'd already decided to live together." She kept shifting her position on the sofa. "It's really helping us to save money. We plan on getting married," she smiled, and then glanced over at the cat again. Her smile quickly faded. "Ed even started taking allergy injections so we could keep Bonsai. But I know Bonsai's mad at me because he has to share me now."

I looked around at the overcrowded room. "Try to see the situation through his eyes," I said, gesturing toward the furniture. "His whole world is upside down now."

I explained to Jill that cats are creatures of habit and that abrupt changes are very frightening to them. Bonsai first had to adjust to Jill's absences from home when she stayed overnight at Ed's apartment, then to the unwelcome permanent appearance of a stranger in the house. Add in all the extra furniture and you've got a cat who doesn't know what's going on. Bonsai had reached the end of his coping ability.

"Has Ed tried to make friends with Bonsai?" I asked.

"Bonsai hissed at him the day he moved in, and, with Ed being allergic, he isn't really all that comfortable around cats," Jill confessed.

"So Bonsai's first serious exposure to Ed was when all his furniture was moved in?"

Jill looked down at her hands. "Yeah," she answered weakly. "But there's something else.

I put down my pen and waited. She stole a quick look at Bonsai and then stared down at her hands again. I saw that same nervousness creep back.

"Jill," I asked softly, "did something happen between Ed and Bonsai?"

At that question, she met my eyes. Usually when clients hesitate, it's because they don't want to admit that they physically punished their cat. I wondered if Ed had hit Bonsai.

"Bonsai . . ." She stopped and put her hands over her mouth. "This is so embarrassing." She became silent for a moment.

"Did Ed punish Bonsai for something?" I asked as carefully as I could.

"Well, sort of." She quickly added, "But he didn't hit him or anything."

"What did Bonsai do that made Ed mad?"

Jill let out a little laugh. "It's really funny when you think about it. Although Ed certainly didn't think it was too amusing." She fidgeted uncomfortably on the sofa. "You see, Bonsai attacked Ed when we were . . . well, we were . . . I thought for sure the bedroom door was closed."

I was getting the picture.

Jill summoned up the courage to spit the words out. "We were in bed, making love." There, she'd said it. "I heard Bonsai growl, but before I could warn Ed, it was too late. He bit Ed on the rear end. He used his claws, too. Ed screamed, grabbed Bonsai, and threw him off the bed. I was screaming, and Ed was screaming. Poor Bonsai, he hissed and ran out of the room as fast as he could. He hid under the kitchen table for the rest of the night." She watched me, waiting for my reaction.

In my most sympathetic tone, I asked, "What happened next?"

"I put antibiotic cream on Ed's wounds. They looked pretty bad. I wanted him to go to the doctor, but he freaked out. He said he wasn't going to a doctor just for some stupid cat bite. But by morning his whole butt was swollen. He could hardly walk. I had to take him to the emergency room."

I glanced over at Bonsai—he hadn't moved from the wicker rocker. His expression was so innocent as he watched us with his large blue eyes.

"The bite on Ed's butt got infected, and he had to stay in the hospital that night," Jill said. "He was so furious with Bonsai. I was afraid he might kill him when he got home."

"What *did* happen when he got home?"

Jill fussed with her hair. "They avoided each other completely. They still do. If Ed's in the living room, then Bonsai stays in the kitchen. When Ed goes into the kitchen, then Bonsai runs into the bedroom. I hate living this way." She leaned toward me. "Can you help us?"

"It would certainly help if Ed could be present during the consultation. Would he be willing to do that?"

Evidently, Ed was so embarrassed by the whole event that he didn't even want to meet me. He had made Jill schedule this afternoon's consultation for a time he wouldn't be home.

"The best way for me to help you is to have Ed here so that I can explain what he needs to do."

Jill closed her eyes and shook her head from side to side. "He'll never agree to it, I know it."

"Would he be willing to do a consultation over the phone?" I asked.

"I guess. Maybe."

We spent the next forty-five minutes going over some behavior-modification exercises for her to practice with Bonsai. Then we arranged a telephone session for later that evening.

Believe it or not, the telephone consultation went very well, although I had a strong hunch Ed was trying to disguise his voice. After a rocky start with him grumbling for a while about how stupid he felt, we got down to business. Basically, my recommendation involved Ed taking over most of the cat-care duties. He was to feed Bonsai and change his litter. He would play with him using interactive toys—the toys providing a safe distance between the cat's teeth and Ed's skin. Also, the interactive toys would help Bonsai associate good things with Ed. In addition, Jill was to be an active member in playtime with Bonsai. I wanted him to see that having Ed move in wouldn't cut down on the amount of attention he'd come to expect from his owner.

Jill and Ed agreed to have a friend store some of Ed's furniture to make life in the apartment a little more comfortable. It would help Bonsai feel as if less of his territory was being violated. I also included instructions for the upcoming move in three months. By then, I was hoping Bonsai and Ed would have bonded enough to make the transition to a new place less traumatic.

At the end of the phone consultation I told them both to call me with any questions or problems. I wanted Ed to remain as active in the behavior-modification process as possible.

"Say, Pam?" Ed asked just before we hung up. "Just out of curiosity, have you ever dealt with a situation like this before? I mean, where a cat bit someone in the . . ." He hesitated. "You know. . . ."

"I did have a client with a similar problem, yes."

Ed let out a big sigh. "You mean somebody else has been bitten on the butt by a cat?" He laughed, obviously glad he was not the first person to have suffered such an indignity.

"Well, it wasn't exactly the man's *butt*—" I started to say, but then Ed interrupted.

"Enough said. I feel a lot better now."

○●○

The Smell

Snuggles, a gray-and-black-striped female tabby, was living a typical indoor/outdoor life in a nice Memphis neighborhood. Her doting owners, retired couple Peg and Frank Foxworth, had installed a pet door in the kitchen so that two-year-old Snuggles could come and go as she pleased.

From the time she was just five months old, Snuggles had proved herself to be quite the hunter. Often she'd charge through the pet door with a mouse, bird, or other small animal clenched in her jaws. Before Peg could stop her, Snuggles would immediately disappear somewhere in the house with her prize. Seconds later she'd reappear (sans prey) and stroll back out through the pet door to begin another hunt. What was so amazing was the speed with which she consumed her treasures.

Of course, Peg and Frank always performed a thorough search of the house, but they could never find so much as a trace of any of

Snuggles's captures. Not a mouse whisker or a bird feather remained. The Foxworths could only assume that Snuggles was devouring her entire prey—feathers, hair, and all.

Frank was not at all happy that his cat was constantly capturing birds. He enjoyed the birds in his yard and was determined to stop Snuggles from getting another one. Peg, on the other hand, was more immediately concerned with the notion the Snuggles was continually bringing rodents inside. Dead or alive, Peg didn't want mice in her home.

Frank's first attempt at ending Snuggles's reign of terror over the backyard wildlife entailed putting a bell on the cat's collar. That worked for maybe two days, after which Snuggles somehow managed to get around this minor handicap and was back to bringing home her usual number of unfortunate victims.

Two bells were put on her collar, but she still remained the envy of every cat in the neighborhood.

Three bells went on the collar. No effect.

Three *bigger* bells were then attached.

Four bells.

Frank finally closed the pet door and started keeping Snuggles inside. She'd be allowed out on a leash two or three times a day so that her owners could keep an eye on her. It seemed to be the only solution.

Snuggles didn't seem to mind being an indoor cat. She played with her toys, lounged in the sun, and kept Peg company as she worked in the kitchen. Being a soft touch for Snuggles's big questioning eyes, Peg began treating her to bits of whatever food she was preparing—a sliver of fruit, a carrot slice, a bit of cooked chicken, or a chunk of cheese. Snuggles would immediately grab the treat that Peg held out to her. And, in true Snuggles fashion, just as she did with her captured prey, she'd dash off to another room with the food in her mouth.

"Maybe she's afraid we'll take it from her," Peg said to Frank one day. "That must be why she always runs off and gulps her food."

"But she never does that with her cat food," Frank replied, pointing to the half-full bowl sitting in the corner of the kitchen.

Peg shrugged and suggested it must be because the mice, birds, and food treats were like prizes, but her cat food was there all the time. Whatever the reason, Peg and Frank happily accepted this small display of eccentricity from Snuggles, since it no longer involved the transport of small animals through their house. If the cat wanted to hide in another room and gulp a piece of cheese, it was fine with the Foxworths.

For a while, life went along happily for Frank, Peg, and Snuggles. Spring passed uneventfully, and then a warm, dry summer came around. It wasn't until fall that things started changing. Snuggles began refusing to use her litter box. A very concerned Peg brought the cat to the veterinarian. She explained that for one week straight now, the cat simply would not use her box. It was kept clean, changed weekly, and had remained in the same spot since Snuggles was a kitten.

The vet examined Snuggles and performed a urinalysis to see if there was evidence of LUTD (lower urinary tract disease). When these tests came back negative, the vet did further testing, but the final result was that Snuggles was a perfectly healthy cat.

When Dr. Walters recommended that Peg contact me, she scowled and said she'd never heard of anything so ridiculous as a feline behaviorist. Her cat certainly didn't need a psychiatrist, and that was all there was to it.

The vet held out my business card and suggested Peg keep it, just in case she changed her mind.

"Don't be silly. My cat doesn't have mental problems," she declared as she walked out the door.

Another week passed. Peg and Frank tried to solve the problem themselves, but the situation was steadily deteriorating. They were constantly cleaning up after Snuggles. Additionally, Peg was very worried now because Snuggles was no longer the happy, playful cat she used to be. Everyone in the house seemed to be in a hopelessly bad mood.

Peg called Dr. Walters and said she might consider talking to the "cat lady."

Three days later, I drove to the Foxworth home for a consultation. They lived on a quiet, shady street in a picture-perfect little home. There was a small yard, neatly maintained, surrounded by carefully manicured shrubs. As I stepped onto the porch I was greeted by an array of soft wind chimes. The large mat in front of the door, with its picture of a smiling cat, issued a friendly welcome.

Peg opened the door, and I entered into a cozy house filled with charm, warmth, and a faint odor of cat urine. She introduced me to her husband, and we all sat down to begin our session.

Peg went over the problem with Snuggles and explained that she'd been rejecting her litter box for the past month. "She began by just occasionally having accidents," Peg said. "Ever since, it's gotten worse."

Frank shook his head. "It was *poof!* Just like that," he said, snapping his fingers. "Right out of the blue she decides she doesn't like her litter box anymore. How do you figure that?"

Peg went on to explain that the box was always kept clean and that she scrubbed it faithfully every week.

"The odd thing is that she's only messing downstairs," Frank said. "She never has an accident anywhere on the second floor."

"Where's the litter box kept?" I asked.

"In the upstairs bathroom," answered Peg.

It was at about this time that the subject of our discussion, Snuggles herself, darted down the stairs and into the living room. She was a small, sleek-looking tabby. Based on the description of her eating habits, I had been expecting to see a large, perhaps even overweight, cat. Snuggles crossed the room and sat at the foot of Peg's chair.

"Have you tried adding a second litter box down here?" I asked Peg.

"No, but we can if you think we need to."

"I think you need to try an additional box on the first floor. But we'll get into that more specifically in a bit."

I asked Peg about Snuggles's personality and habits. Peg repeated the stories of Snuggles's former rapid-fire—yet secretive—ingestion of prey and her current need to gulp her treats in private. Interesting, I thought. The cat did appear a bit high-strung.

After I had asked all of my questions and spent time with Snuggles, I began a tour of the house. The rest of the main floor was very much like the living room—cozy, inviting, and very neat. I could picture many wonderful home-cooked meals begin prepared in the kitchen and lots of enjoyable evenings sitting by the fire in the den, surrounded by children and grandkids. It was as we ascended the stairs to the second floor that my feelings began to take a dramatic turn.

In appearance, the upstairs appeared to be just as nice as the main floor. The first sign that something was amiss was the overpowering floral smell. It practically knocked me down as I reached the top of the stairs. It was a sickeningly sweet smell, verging on being suffocating. I looked around, and there were automatic room fresheners everywhere. The bathroom alone had three—one on the top of the toilet and two on the counter, just inches away from each other. But that wasn't all. Electric plug-in fresheners could be seen sticking out of every wall outlet. Mounds of potpourri sat in huge bowls everywhere I looked. At first I thought the air fresheners were an attempt to mask any urine smell, but then I remembered that Frank had said Snuggles never messed upstairs.

As I walked around, I wondered if the Foxworths had been this heavy-handed with the room fresheners before Snuggles's accidents. If so, this may very well have brought on the problem. No wonder the cat was using the downstairs carpet. It was the only place she could recognize her own scent.

I tried to detect any scent of cat urine, but even my good nose was no match for the arsenal of aromatics Peg had put out. There was something in the air, but I couldn't quite figure it out. It wasn't urine, but it was *something.* It smelled kind of spoiled. No, worse than that. It was very weird. Mingled in with the suffocating sweet smell was the smell of something *rotten.*

Since my sense of smell wasn't able to help me detect any traces of urine, I pulled out my black light, which causes urine stains to fluoresce on walls and carpeting. I checked around, dreading having to go down on my knees—my sudden repeated sneezing told me there had to be numerous powdered deodorizers on the carpets.

So far, the black light had detected no visible urine stains upstairs. Even though the Foxworths had told me Snuggles didn't mess there, I just wanted to double-check. The only room I had yet to examine was the last guest bedroom.

As I approached the room, the thick, sweet smell grew in intensity. It was almost unbearable. I looked back at the Foxworths, feeling sure they'd say something. Anything. *Can't you smell this?* I screamed in my head. *Please give me a rational explanation as to why the upstairs of your home smells like a combination of a morgue and an overstocked florist shop.* But they both just looked at me with blank expressions.

I have to admit that for one brief moment (okay, maybe two or three) I had the frightening thought that perhaps this sweet old gray-haired couple had murdered someone and were attempting to disguise the smell of the decaying corpse. Farfetched, I realize, but by this time, the lack of fresh air was getting to my brain. Anyway, who buries a body in an upstairs bedroom?

We stood in the guest room—the smell was definitely originating from here. I had to find out what was going on, so I turned to Peg.

"I notice you have several air fresheners upstairs here," I said casually, attempting to be subtle. "In the bathroom, there are three right near the litter box. I'm afraid that Snuggles's problem with the box could be connected to the perfumed scent of the air fresheners.

It's probably too overpowering for her. She needs to be able to smell her own familiar scent." I took a small breath (a large one probably would have killed me). "Did you always have these air fresheners?"

"Why, no," Peg answered innocently. "We were trying to get rid of the smell."

Ah, yes, the *smell!*

"The smell of urine?" I asked, knowing that couldn't be the smell she was referring to, but the image of the nice old Foxworths as ax murderers had just flashed through my oxygen-deprived brain again.

Frank shook his head. "No, that *bad* smell. We think it's from when we had the walls painted a few months ago. It really stinks."

That smell had nothing to do with the paint. Something or someone had died in here. "The smell is much worse in this room," I said to them. "Was it the only one painted?"

"No," Peg answered. "They did the whole upstairs, but for some reason it smells worse in here. It didn't seem that bad at first, but it kept getting stronger as the paint dried."

At that moment, the things Peg had told me earlier about the way Snuggles would race in the house with a mouse and disappear to eat it flashed through my mind. I was trying to fit the pieces of this puzzle together. Peg and Frank had said they were amazed at the speed with which Snuggles devoured her prey, and that no traces were ever found. I had a hunch that somewhere in this room there was a stash of dead mice, and who knew what else. It made sense. Now, how did I tactfully tell Snuggles's owners that they probably had a pile of dead animals (or at least pieces of animals) in their pretty little blue-and-white bedroom with the dainty lace curtains?

I decided to just be direct.

"What?!" Peg was understandably shocked. "I clean these rooms thoroughly each week. Don't you think I'd notice a bunch of dead mice in the corner?"

Although I try never to offend my clients, I feared I was walking a thin line here. But I knew I was right—I just needed evidence. And, based on the stench in this room, I was sure the evidence was very close by.

With perfect timing, Snuggles walked into the room. She watched us warily for a moment and then dove under the bed.

I was listening to Peg rationalize why she was convinced it was the paint causing this horrible smell when I thought I heard Snuggles growling. I looked at Peg and put my finger to my lips. "Sssh," I said quietly. "Listen."

The sound was quite audible to all three of us.

"Is that Snuggles?" Frank asked. "Why is she growling like that?"

I had a very strong suspicion that I knew the answer, but before I risked offending my clients again, I needed to do something. I excused myself and went downstairs to the living room to get one of my cat toys.

Back in the bedroom, I sat on the carpet—sneezing several times in the process—and dangled the toy near the door.

"What are you doing?" asked Peg with more than a hint of impatience.

"I need Snuggles out of the room for a minute. Please bear with me."

Within seconds, Snuggles, ever the mighty hunter, peeked out from under the bed and pounced on the toy. I kept leading her farther out into the hall. When we had cleared the doorway, I reached into my pocket, fished out a furry mouse (fake, that is) rubbed with catnip, and gently tossed it into the other bedroom. Snuggles ran in after it.

With the cat now distracted in the other room, I went back into the odoriferous room, closed the door, and turned to a puzzled-looking Peg and Frank. "Please bear with me a moment longer. I have a hunch I may know where this smell is coming from."

Peg frowned. "I've already told you—"

Frank interrupted her. "Please, Peg, hear her out."

"Thank you," I said meekly. I knew I'd better be right. I had already offended them once; I didn't want to try for twice.

Bending down, I lifted up the bedspread and looked under the bed. The smell just about knocked me over. I held my breath. Just as I had suspected, the cheesecloth under the box spring had a huge hole torn in it. I stood back up, took a few breaths of the rotten, perfumed air, and asked Frank to take a look under the bed.

"Did the damn cat do that?" he asked as he straightened back up.

"What'd she do?" Peg asked.

"Tore a hole in the box spring," he answered. "Damn cat."

"Did you notice anything else?" I asked.

"What do you mean?"

I pointed to my nose. Frank bent down again and checked under the bed a second time.

"Oh, Lord, it stinks under here!" he exclaimed.

Thank you. "I think Snuggles has been stashing her prey in the bed," I said. "That's why she was growling at us. We need to turn the box spring over and check inside. But first, can we please open a window?"

Without hesitation, Frank started tearing the bedspread off and pulling the mattress aside. Together we grabbed the box spring and turned it over. Peg opened a window.

"Get me a flashlight," Frank said to Peg, who was standing by the window with her hand over her mouth. "Damn it, Peg, get a flashlight."

She quickly turned and dashed out of the room. Moments later she was back, holding a large flashlight.

I looked at Frank. "Do you want me to check?"

He hesitated, then took the flashlight from his wife. "No, I can do it." He peered inside the hole.

"Oh, no!" Frank said, suddenly moving away from the bed.

I took the flashlight, held my breath, and looked inside the box spring. It was bad—real bad. Snuggles had been using the box spring as a rodent cemetery; it was full of decaying mice, birds, chipmunks, and squirrels. In addition, the cat had also apparently stashed away every morsel of treat food Peg had ever given to her. There were chunks of moldy cheese, spoiled meat, and many more unrecognizable items.

"What is it?" Peg pleaded.

"Don't look," warned Frank. "The cat's been hiding those animals she brought home."

"All of them? In there?" she pointed. "Oh, Lord."

Frank and I carried the box spring down the stairs and out to the backyard. He came back into the house and was immediately on the phone with someone about hauling the nasty thing away.

Peg fixed us all iced tea, which we drank on the front porch in the much-needed fresh air. I explained to them that their efforts to cover up the unpleasant odor with air fresheners had created an environment in which Snuggles was no longer surrounded by her own familiar scent. Peg agreed to remove all the fresheners, thoroughly clean the upstairs, and air the place out. Once that was done, they would begin play sessions with Snuggles to help her become comfortable and happy about being upstairs again.

We also talked about putting an additional litter box in the downstairs bathroom. And, because there was still a trace of urine odor downstairs, I suggested the Foxworths have the carpet professionally cleaned. I recommended that they keep a bottle of odor neutralizer in the house at all times, for any future accidents. We needed to completely neutralize the urine odor so that Snuggles wouldn't smell it and continue using those areas as her litter box.

Play sessions were to be conducted in the areas that Snuggles had used for urination. The objective was to change those areas in the cat's mind from a litter box area to a nest area. Peg could also place some food bowls in those spots. Cats don't eliminate in the nest, where they sleep, eat, and play. In the wild, they eliminate away from the nest so they won't attract predators. I wanted Snuggles to view her home as her nest again.

We went over a few more specific behavior-modification exercises, and then I talked about how the Foxworths could avoid having another boxspring turned into a rotting shrine for Snuggles's hunting trophies. Peg and Frank had already been keeping Snuggles indoors, so that ended her capture of outside prey, but the food treats needed to stop also. Normally, I see food-hoarding more in multipet households, although there are some solitary cats who have a strong desire to store their treasures in a little hideaway. However, I'd never seen it to this degree.

I explained that even if Snuggles didn't have the urge to hoard her goodies, table scraps interfere with a cat's normal nutrition. Snuggles would be much better off if she stuck with her usual cat food. Peg agreed.

Finally, just to be on the safe side, I recommended that they cover the bottom of all their box springs with fitted sheets.

When the session ended, Peg thanked me, then excused herself to go upstairs and begin cleaning. "It's amazing how we just got used to that smell," she said as she started up the stairs.

Frank walked me out to my car. As we stood in the driveway saying our good-byes, I caught sight of the boxspring lying in the backyard. The thought of a guest sleeping on that bed, on top of all those dead animals, made my stomach start to churn.

Suddenly, the memory of that sickeningly sweet smell overtook me. I quickly wished the Foxworths good luck, got into my car, and drove home with the windows wide open.

The Cat Who Hated Sundays

The aggression problem that Carolyn Tinnerly called me about was notable in that it occurred only once a week—and always on schedule. Mambo King—her three-year-old, cream-colored, male Persian—had attacked her repeatedly one Sunday, and had taken to ambushing her every Sunday since. Carolyn's call to me came after the sixth Sunday attack.

"He's in a terrible mood every Sunday," she said. "He's a wonderfully sweet and affectionate cat every other day, but he turns into this ferocious tiger on Sundays. I just don't get it."

These attacks had left Carolyn with bites, scratches, and torn clothes. For the past few Sundays, she had tried her best to avoid Mambo King. Often, she'd close the door to whatever room she was in and check to make sure the coast was clear before moving from one room to the next.

"I can't have any guests over on Sunday unless I know which room he's in and can lock him inside," Carolyn said sadly.

We agreed on an appointment for the following Sunday.

Carolyn Tinnerly's house was very large, and the surrounding property was beautifully landscaped. After parking in the driveway, I sat in the car for a minute, reflecting on the painstaking work that must have gone into creating such a lovely yard.

I heard my name and looked toward the side of the house. Carolyn was standing at the back door and waving me in.

"Come through the kitchen door," she said as I got out of my car. "It'll be safer. Mambo spends most of his time in the living room on Sundays."

She led me into the kitchen, where I put down my things. As I was removing my jacket, I heard the faint but-all-too-familiar sound of a cat growling in another room. Carolyn heard it also and nodded her head.

"That's him," she confirmed with a weak smile.

As if on cue, the cat appeared in the doorway. He greeted me with a long, low growl. His body looked tense and poised, ready to spring at any moment. I was treated to three hisses, and then he suddenly took off, flying back toward the front of the house.

A heavy sigh came from Carolyn. I believe she'd been holding her breath during Mambo King's entire appearance. I have to admit that I breathed a slight sigh of relief myself.

"He's like this every Sunday," she sad as we sat at the kitchen table. "The rest of the week he's the sweet cat I've always known."

Mambo King's most recent vet visit revealed him to be in excellent health. He had never had a litter box problem, loved people, was affectionate, slept in his owner's bed at night, and during the day had the run of the house. That was the other six days of the week. But when Sunday came around, Mambo King chose to stay in the front part of the house—which consisted of the living room, dining room, and home office—hissing and growling most of the day. But most upsetting were his attacks on Carolyn.

"This was the latest," she said, and she raised the cuff of her pants to reveal several deep scratches. "I've had to be on antibiotics this time around." She smoothed her cuff and leaned back in the chair. Her eyes filled with tears. "I love him," she said. "I don't want to have him put to sleep."

I put down my pen. "Let's see if we can find some answers first," I offered.

Carolyn ran through her routine for each day, but she could think of no changes that might have triggered the attacks. She had lived in the same house since adopting Mambo. There had been no new furniture recently, no job changes, not even a visit from any unwelcome houseguests. "Mambo loves company," she added. "Even if I had had a house full of people, he wouldn't have objected." She looked sadly toward the living room. "Except on Sundays."

Sitting at the kitchen table, I had Carolyn go through her Sunday routine as specifically as she could remember. The few rituals that were exclusive to Sundays were things she said she'd done all of Mambo's life, without any previous problems.

My first thought as to the possible cause for Mambo's weekly aggression was the appearance of an outdoor cat. A cat who, for whatever reason, came around once a week.

At the mention of another cat, Carolyn nodded. "There's a big, scruffy-looking white cat who comes by to check my bird feeder," she said. "But he's been around on other days, not just Sundays."

According to Carolyn, Mambo's reaction to the white cat had always been negative—a hiss at the window or an occasional growl. But these episodes never resulted in redirected aggression toward his owner, except on Sundays.

I began my tour through the house, checking windows to see which ones had a view of the bird feeder. My plan was to partially block Mambo's view from these windows so that he wouldn't see the white cat.

"But before I block the windows," I said, "why don't I first visit with Mambo?"

Mambo King was waiting for us in a corner of the living room;

he had positioned himself near the floor-to-ceiling windows. I noticed that his facial expression and body language displayed all the signs of a highly agitated cat. He lowered his head and flattened his ears. From his crouched position, he issued a serious growl. This was accompanied by a series of hisses.

"If you're afraid, you can wait for me in the kitchen," I said to Carolyn, while keeping my eyes on Mambo.

Carolyn reached across the coffee table and grabbed a plastic squirt bottle. "I've been using this whenever he comes near me," she explained. "I have one in every room." She placed the bottle back on the table and walked out of the room.

It saddened me that the relationship between cat and owner had come to this, but I certainly understood her concern for her safety.

Mambo watched me cautiously. I stood in one spot for quite a while and then slowly lowered myself to the floor to get in a comfortable position for our session. Within a few minutes it seemed as if Mambo came to the conclusion that I wasn't posing any immediate threat, and he began walking toward the center of the living room, where I was sitting. He appeared to be somewhat less nervous as he neared me. But when he passed the large front window, with its sunny window seat and array of plants, he stopped abruptly. Whipping his head to the side, he hissed in the direction of the window, then circled around the potted plants. Adding a growl to his emotional display, he cast an extremely hostile look in my direction. I sat motionless.

Mambo began pacing back and forth in front of the window, getting angrier with each pass. Without moving any part of my body except my eyes, I tried to see if there was indeed a cat outside. Mambo's agitation increased. Suddenly he bolted from behind the largest plant, heading directly toward me. I had just enough time to roll to the side and miss his oncoming attack. Luckily, he chose to keep going and disappeared down the hall. I immediately went to the window. No sign of any cat.

As soon as I approached the window, I caught the unmistakable whiff of cat urine. Now, when you've been doing this work as long as I have, you develop certain strengths. One of my well-developed strengths is my nose. To the casual observer, it's an ordinary-looking nose, but don't let that fool you. My nose can detect even a tiny amount of cat urine. Running into cat urine in unexpected places is a daily occurrence for me. At first I thought the smell must have been coming from the carpet, but as I knelt down, the odor seemed to be limited to the plants. I touched the soil in one of the pots, and it was wet. I sniffed again and, sure enough, there was the faint smell of urine. Mambo was spraying the plants. I headed back to the kitchen to talk to Carolyn.

"I've never seen him spray," she said, very surprised.

"But I do smell the urine," I explained.

We went back into the living room so I could do a closer inspection. The urine was definitely coming from the soil surrounding the potted plants. As I looked closer, I noticed the soil in the pots wasn't the kind I normally see with indoor plants. This soil looked and felt more like ordinary outdoor soil.

"Did you buy packaged potting soil for these plants?" I asked, sure that I already knew the answer.

"No, I used soil from my garden," she replied. "Why?"

Maybe Mambo hadn't sprayed the plants after all. "I think the soil might have been contaminated by another cat's urine," I said, watching Carolyn wrinkle her nose in disgust. "Probably the cat who comes around here so often."

"He does spray the redwood chair on the front porch," Carolyn added.

"I'll bet you water your plants every Sunday, right?"

"Yes."

The puzzle was fitting together.

"There's your reason for the weekly aggression," I explained. "Every time you water, it reactivates the urine odor. Mambo smells the intruder's urine and it sparks his territorial instincts."

With another feline mystery solved, I helped Carolyn remove the plants from her house. We put them in the garage, where she could safely do the repotting without agitating Mambo further.

Back in the house I laid out a specific behavior-modification plan to help Carolyn reestablish her close relationship with Mambo. Part of that plan included using an interactive toy to play with him in the area where the plants had been. I wanted Mambo to feel safe again there and to start associating that part of the house with good things.

If Mambo again became agitated after seeing the outdoor cat, I instructed Carolyn to block his view of the yard by attaching a poster board to the bottom half of the window.

A very important assignment for Carolyn was to find out who owned the outdoor cat. If she was unable to locate an owner, the local humane shelter would have to be contacted.

Since the white cat was using Carolyn's garden as a litter box, I strongly advised her to wear gloves when gardening to prevent possible toxoplasmosis, a disease that can be passed to humans through contaminated soil. Also, Carolyn was to remain aware when gardening that the urine smell could get on her shoes or pants; I suggested she immediately change her clothes after working in the yard.

Finally, when the plants were repotted using packaged soil, Carolyn was to place them in a different window to help break the negative association.

Two days after my house call, Carolyn phoned to say the white cat had reappeared. She had taken the cat to the shelter and placed an ad in the paper. Carolyn paid for his FeLV testing and donated money toward vaccinations and neutering. He was a nice cat, and the shelter felt confident he would be adopted if no one claimed him.

The following Sunday evening I received another call from Carolyn to say there had been absolutely no aggression from Mambo King. He was his old sweet self again. The repotted plants had been placed in another part of the house without any negative reaction from the cat.

Carolyn was also able to remove all the water-filled squirt bottles from around her home. She had thrown them all in the garbage, she said, because they held a negative association for both her and Mambo. Carolyn would be able once again to enjoy her old Sunday ritual of sleeping late, lounging on the living room carpet to read the paper, and having friends over.

During a call to the shelter I was told that the stray cat had been neutered and adopted. Three months later I received a call from his new owners, asking for help with a spraying problem. I made a house call, and he's been doing much better; as of this writing, there haven't been any more spraying episodes.

Carolyn and Mambo King are also doing well, but Carolyn won't water her plants on Sundays anymore. Wednesday is the new day.

"Just to be on the safe side," she says.

○ ● ○

Cassie's Gift

I first met Cassie four years ago, under very distressing conditions, when her owners phoned me to ask for help with their melancholy cat. It seemed that Cassie's companion, Ginger, a seventeen-year-old Siamese, had died several weeks earlier. Marie Donaldson, one of Cassie's owners, said that Cassie remained severely depressed; she was barely eating, and now spent most of her time sleeping on Ginger's favorite chair. The Donaldsons, who were still dealing with their own grief, were hoping that Cassie would eventually adjust to life without Ginger, but her despair seemed only to grow deeper.

When I made my house call to the Donaldson home, my heart broke for all three of them. They were having an extremely difficult time coping with the loss of Ginger.

What I learned during that session was how close the family was. Marie and Paul Donaldson had been totally devoted to their cats, and, in return, the two cats had given them an abundance of love. There were cat trees and toys all over the house. Catnip plants grew on the windowsills, and Paul's homemade scratching posts could be found in almost every room.

Cassie had entered their lives unexpectedly when Paul found the tiny kitten abandoned in the parking garage near his office. Thin, dirty, and sick, she was sitting in the middle of an empty parking space. With barely enough strength to let out a faint meow, she called to Paul. The frail kitten was brought home in Paul's jacket pocket. That evening, Marie took her to the vet, where she was diagnosed with a respiratory infection.

For the next fourteen days, the little cream-colored kitten was nursed back to health in the upstairs guest bedroom. On her return visit to the vet, she was declared healthy and received the first of her vaccinations.

From the moment the tiny Cassie met the Donaldsons' resident cat, Ginger, it was love. Cassie followed Ginger everywhere. When Ginger went to the litter box, Cassie was right behind her. When Ginger sat by the window to watch the birds, her little feline shadow was at her side. For her part, Ginger seemed to know how desperately Cassie needed a mother, so she did her best to fill those shoes. She groomed her little companion, shared her toys, and let Cassie curl up right next to her at bedtime. The two became inseparable.

As I sat and listened to Marie and Paul describe life with their cats, I watched Cassie sitting in the corner. Every time a tearful Marie mentioned either cat's name, Cassie, not used to hearing that tone of voice, looked unsure of whether to come over to Marie or stay away. She finally did walk over to Marie hesitantly. She was immediately clutched so tightly by her owner that she very quickly struggled to get free.

"Cassie, I need you," Marie pleaded as the cat darted back to the safety of her corner.

I felt the pain everyone was going through, but I needed to explain to Marie that her clutching at Cassie, combined with the despair in her voice, was creating fear and confusion in the cat. In Cassie's mind, not only was her beloved friend gone, but her trusted owners were suddenly acting differently toward her. Everything she was used to had changed.

"Cassie's mourning the loss of Ginger. But what will comfort her the most is to have her normal routines continue unchanged," I explained. "She needs all the usual playtime she used to have. And she needs her owners to be the way she remembers them." I got up from my chair and sat down on the sofa next to Marie. "Even though you're experiencing sorrow now, for Cassie's sake, please try to keep things as normal as possible for her. Cats are like sponges; they absorb whatever we're feeling."

I suggested behavior-modification techniques to help all three of them during this difficult adjustment period. I also provided them with the number of a grief-support hotline specifically for people struggling with the death of a pet. "The people involved in grief support understand what an important family member a pet is," I told them, "and how deep the sorrow is when the pet dies."

During the next several weeks I kept in close contact with the Donaldsons and made a number of visits to their home. Their relationship with Cassie was getting back on track again, and life began returning to normal. I stayed in contact with them over the next couple of years via regular phone calls.

Two years later Cassie was taken to the hospital because Marie had noticed she'd lost a significant amount of weight. The veterinarian diagnosed renal failure. At the age of sixteen, Cassie's kidneys were beginning to shut down. Luckily, the vet said, the disease was in a very early stage, and there were steps that could be taken to help Cassie.

Cassie stayed in the hospital for a few days. When Marie came to pick her up, she was given a prescription food, which she was told Cassie would need to remain on for the rest of her life. After some additional instructions, Marie took her cat home.

Within a few days, Cassie seemed to be her old self again. The Donaldsons kept her on the prescription food and brought her back to the vet regularly for monitoring. Cassie's condition didn't seem to be deteriorating at a rapid rate until several months later, when she took a devastating turn for the worse. She barely had an appetite. Then one day Cassie refused all food and water and wouldn't move from her favorite chair. Marie and Paul brought her into the animal hospital that evening, where she was put on IV fluids. The vet knew the past few weeks had been difficult for everyone. The only way Marie had been able to get food into Cassie was to hand-feed her. She was also keeping the cat hydrated with subcutaneous fluid therapy. Now Cassie was refusing the hand-fed food.

After being hospitalized for five days, Cassie's condition improved and she was allowed to go home.

Cassie had her ups and downs over the next few months, until finally it was apparent that her struggle was taking too much out of her. Marie and Paul talked with the vet, and it was decided that Cassie was enduring needless suffering. Sadly, an appointment was made for them to bring Cassie in the next morning to be euthanized.

It was extremely difficult for the Donaldsons to get through the evening. They stayed by Cassie's side, trying to keep her comfortable. Cassie was restless and kept looking out the window. At one point she began meowing loudly and scratching at the glass. Paul looked out into the darkness but couldn't see anything. Marie feared that Cassie was reacting to their feelings about putting her to sleep.

All evening Cassie couldn't be comforted. Even in her weakened state, she tried to scratch at the window. Marie closed the curtains, but Cassie still wouldn't let up.

"Go see if there's something out there," a distraught Marie instructed her husband.

Paul put his coat on, dug a flashlight out of the kitchen drawer, and headed outside into the chilly darkness. As he walked around the yard his flashlight beam caught something. He inched closer. He had to get down on his knees to see, but in a pile of leaves, under a bush, was a cat. At first Paul thought the cat was dead, since she was motionless and on her side. Then he heard a painful-sounding meow, and the cat struggled to get up. It was then that Paul saw the cat wasn't alone. Underneath her lay three tiny kittens. Paul quietly backed away so he wouldn't frighten the cat, then raced inside to get Marie.

I was called. Paul led me to the backyard to see the mama cat. She looked frightfully thin and weak. We could see that she was trying to shield her kittens from the cold, but we were sure they would die if we didn't get them inside fast.

I went back to the house and got the carrying case I'd brought. (I never travel without a cat carrier. I may go out without combing my hair or locking my door, but I never forget my cat carrier.) I asked Marie for some canned cat food, and then heated a small plate of the food in the microwave so the aroma would be more enticing.

"I'll try to get the cat to come over and eat," I told Paul and Marie. "You both stay out of sight. When I get her in the carrier, you get the kittens." I put on my animal-handling gloves in case the cat was difficult to grasp; but she looked so weak, I didn't feel she posed much of a threat.

Outside, I slowly approached the mama cat and put the cat food down on the grass beside me. I watched her nose go up in the air as she caught the aroma. She meowed with interest. Picking up the food, I came a little closer and set it down again. I knelt down and started talking to her. She looked at the food, then struggled to get up. Her kittens meowed and squirmed as she moved away from them and toward me. She was wobbling unsteadily. I extended my hand for her approval sniff, and she allowed me to pet her. As my hand ran down her thin back, she began to purr and then dove into the food. After I let her eat a bit, I picked her up and put her in the carrier. Marie and Paul were right behind me as they raced over to scoop up the kittens.

Inside the house, Marie had set up one of the bedrooms for the cat. There was a towel-lined box in the closet, a comfortable place for her to nurse her kittens in private. Marie had also created a makeshift litter box out of a plastic dishpan. In the closet was food and water for the hungry mother.

I waited until Marie and Paul had placed the kittens in the box, and then I opened the carrier. The cat went immediately to her babies. As soon as she got in position, her kittens nestled right up against her.

The Donaldsons went back downstairs to sit with Cassie while I stayed with the mama cat. I pushed the dish of food over to her and she ate hungrily.

When the dish was empty, the cat closed her eyes and rubbed her face against my hand. I checked each of the kittens, and they all seemed to be okay. Within a few moments the mother lowered her head and fell asleep. The poor thing was completely exhausted.

I went downstairs to check on Marie and Paul. They were with Cassie, who was now quietly lying on her cushion.

"She's not restless anymore," Paul said.

"She was trying to tell us about the cat outside," Marie said tearfully as she stroked her cat.

"Cassie saved that cat and her kittens," I told them. "With the temperature dropping the way it is, the kittens definitely would not have made it. And their mother is so weak, she wouldn't have had the strength to move them."

Marie asked me to spend the night so that they could stay with Cassie and I could keep watch over the new family. I slept in my clothes that night and made sure the mama cat took breaks from her duties periodically in order to eat.

The next morning was the day Cassie's euthanasia was scheduled. We gathered up the kittens and their mother to take with us for the vet to examine. The Donaldsons didn't know what they were going to do with the mother and her kittens. They were hoping they could board them at the vet until we came up with a plan.

Paul bundled Cassie in a blanket and placed her in Marie's arms. After helping Marie get into the car, he assisted me with the kittens and their mother. We placed the box on the front seat of my car, and I followed the Donaldsons to their vet.

Once at the hospital, I handed the box of cats to an animal technician. Marie was still holding Cassie. As the technician was walking away, the mother cat stuck her head over the top of the box and looked around. When she spotted Cassie in Marie's arms, she meowed. It was a very soft, gentle meow. Cassie then lifted her head, looked over her shoulder at the cat, and meowed in return. It was an identical-sounding meow. I know the grateful mother was saying, "Thank you," and Cassie replied, "You're welcome."

Cassie died peacefully in Marie's arms that morning, by way of merciful euthanasia. She was eighteen years old. And although the vet agreed to keep the mother cat and kittens at the hospital for a while, the Donaldsons knew that Cassie would have wanted the new family to go back home with them, even if it was only for the time being.

Two weeks later the kittens were doing fine and growing rapidly. The mother, now named Lyla, was getting stronger and healthier by the day. The Donaldsons deeply missed their Cassie, but they were so proud of the unselfish concern she had shown for another cat in trouble. Indeed, Cassie had given a very precious gift to the mama cat and her kittens. She passed along to them her loving family, the Donaldsons.

By the way, Marie and Paul decided to keep Lyla *and* her three kittens, whom they named Miracle, Sandy, and Little Cass.

Why Henry Isn't Himself These Days

I like to think that I know my cats very well. I know their likes and dislikes, I'm in tune with their body language, and they're certainly in tune with mine. If any of my cats don't seem to be their usual selves, I know right away and can immediately begin trying to find the cause. For owners Sharon and Steve Ruhl, though, when their cat wasn't acting like himself, it was time for a house call.

It all began when Henry, a five-year-old black mixed breed cat, who had never showed any interest in going outdoors, slipped outside when no one was looking.

The Ruhls lived in an old ranch-style house that they were in the time-consuming process of renovating. Floors squeaked, doors creaked, and the few windows that hadn't been painted shut had old, ill-fitting screens and storm windows. In the spring, Sharon and Steven opened the windows (with sturdiest of the screens) just enough to let in fresh air, but not enough to pose a risk to Henry.

The Ruhls were having new wallpaper hung in several rooms, and one evening, as the workers were leaving, Sharon couldn't locate Henry but wasn't too concerned, thinking that he must've just found his own hiding spot to the avoid the day's commotion.

It was not until after the workers were gone that Sharon noticed the screen had been popped out of one of the windows. Feeling the panic rush over her, Sharon checked each room, but there was no sign of Henry.

Sharon combed every inch of the house and then began searching outdoors, repeatedly calling out Henry's name. As it became more obvious that Henry wasn't in the immediate area, Sharon began to cry. She ran into the house and paged her husband, who then left work early to join his wife in the search. The couple spent a long night looking for Henry until Steve convinced his frantic wife that they needed to get some sleep.

"Henry will probably show up in the morning," Steve assured her. "He'll want breakfast so we'll probably find him sitting by the back door."

Morning came and went with no sign of Henry.

The Ruhls put up posters, called all of the neighbors, informed the local shelter, and spent every evening calling for Henry in the backyard. This went on for weeks.

Sharon and Steve were about to give up hope. After five weeks of searching, they doubted they'd ever see Henry again. Steve only wished that a kind person had perhaps picked up Henry and would provide him with a loving home. Sharon blamed herself for everything because she hadn't locked the cat up before the workers arrived. Every time she looked at the newly papered walls, she was reminded of that horrible day.

One morning, three months after Henry's disappearance, Sharon was taking out the trash when she noticed movement in the bushes near the garage. Quietly, she placed the bag of trash on the ground and knelt down to get a better look. She fixed her eyes on the bush, trying to determine what kind of animal it was. Suddenly, a set of eyes was staring back at her from between the branches. Tears began to run down Sharon's cheeks.

"Henry!" she began to scream, then abruptly cupped her hand over her mouth for fear of scaring him off. Using all of the restraint she could gather, Sharon sat down on the sidewalk and quietly spoke to her cat.

"Come to Mommy, Henry," she whispered while fighting the overwhelming urge to run over and grab him.

The cat remained where he was, but continued to watch her with interest.

"Henry," Sharon pleaded, "come here, baby."

The cat started to inch his way out of the bush and Sharon extended her arms, ready to grab him—but he stopped just out of her reach. She could now see how thin and straggly he looked. Ever so slowly, she unfastened the twist tie from around the trash bag and began searching inside for some leftover morsels. She pulled out some shriveled up vegetables and rotting fruit cores. She didn't think that would be very appealing, even to a hungry cat, so she continued to dig through the bag. Finally her fingers landed on something—a fish, or at least parts of one. Thank goodness she had decided not to finish all of her broiled snapper the previous night. Sharon pulled the smelly remains out of the bag and placed it on the sidewalk in front of her. If the morning breeze continued to cooperate it would only be seconds before the cat's nose got the message.

The snapper worked its magic and the frail-looking cat tentatively approached the food. Sharon watched and waited for the perfect moment to grab the cat. As soon as Henry got close enough, she snatched him up into her arms and ran into the house. Once inside the kitchen, while still holding Henry in her arms, she opened a can of cat food and put it on the floor. Then she gently placed her cat in front of it. With just a slight initial hesitation, Henry made short work of the meal. Sharon's tears continued to fall. Henry had come home.

That night, when Steve returned from work, the tearful reunion continued. Aside from looking dirty and quite thin, Henry seemed happy to be home. He went from room to room, checking everything out, until he finally fell asleep on top of the refrigerator. It was not a place he used to go; in fact, he had been trained not to jump onto anything in the kitchen, but Sharon decided to overlook any rule-breaking for the time being.

The following morning, Henry not only enjoyed his own breakfast, but much of Steve's, as well. When Sharon placed a plate of pancakes on the table and then called to Steve from the bottom of the stairs, Henry wasted no time in seizing the opportunity to help himself. Sharon returned to find half of the pancakes gone and several tooth marks embedded in the butter. Apparently, Henry was now quite accustomed to helping himself.

After breakfast, Sharon took Henry to the vet for a checkup. Henry's usual veterinarian was on vacation and so a relief vet, someone Sharon wasn't familiar with, examined the cat. Aside from being dirty and skinny, Henry seemed to be in good health. Sharon was thrilled.

Over the next couple of days, Sharon, Steve, and Henry began the process of getting reacquainted with one another. Henry's outdoor adventure had resulted in some changes to his personality. He tended to keep to himself during the day, preferring to sleep on the top of the refrigerator or under the guest-room bed. At night, it was a different story, though. This was the time when he chose to play. Sharon and Steve would hear all kinds of noises and bangs after they got into bed. Upon rising in the morning, several small objects could be found on the floor. Sharon remained unconcerned because, after all, Henry had been through a frightening experience. She was willing to put up with a few minor inconveniences.

Another change in Henry had to do with his eating habits. In the past, he had been a picky eater, barely nibbling at his cat food. Now, he not only devoured his own meal, but if Sharon wasn't watching, he'd jump onto the counter and grab anything edible—even snatching food right off the stove. He had also begun raiding the kitchen trash can on a regular basis until Steve finally decided to relocate it in the cabinet under the sink.

"So that's where last night's dessert went," Sharon said to Steve one morning as she was vacuuming and discovered several broken bits of cookies under the couch. "Henry must not have been a very good mouser. He probably learned to live out of garbage cans."

"He'll get over it in time," Steve answered over the noise of the vacuum trying to suck up the pieces of cookies. "In the meantime, we'll just have to be more careful."

Sharon nodded in agreement, but inside, the thought of how Henry must have had to struggle made her so sad. She longed to hold him in her arms and cuddle him the way she used to, but Henry didn't seem ready for that yet. Whenever she tried to pick him up, he ran out of the room. Time . . . it would just take more time, Sharon told herself.

By the third day, though, Henry's readjustment process took a major step backward. The cat began to spray. It started with just a small amount of urine sprayed on the wall in the bathroom, next to the litter box. Sharon took this in stride, figuring that it was still too soon to expect much from Henry. But by the next morning, it was obvious that something had to be done. When Sharon and Steve awoke, they found Henry had sprayed his cat tree in the bedroom. They didn't understand it, because the tree used to be his favorite place to nap before he disappeared.

After cleaning up the mess, Sharon headed toward the kitchen to make some much-needed coffee, feeling some relief from the fact that at least Henry could no longer raid the trash can, since they'd placed it in the cabinet. She didn't want to have to clean up spilled garbage after just having cleaned up all that cat urine. "Thank goodness we at least solved one problem," she said to Henry as he darted past her in the hallway and disappeared under the guest-room bed.

Of course, raiding the trash can was a piece of cake (so to speak) compared to what Henry had in store for Sharon as she rounded the corner from the hallway and stepped into the kitchen. At first, Sharon thought there must have been some kind of leak, but as she walked closer, the reality hit her as powerfully as the smell. Her stove, countertops, and cabinets had been sprayed. Most of it had dried, but the cabinet area over the stove was still dripping urine. Henry must have begun spraying the previous night and continued through the morning.

Sharon managed to call out Steve's name before sinking down in the nearest kitchen chair with her face in her hands. Steve was unable to hear her at the moment, though, because he was bending down in the bedroom closet, reaching for a pair of shoes from the neatly arranged row along the wall. As he bent closer he noticed that some of the shoes in the corner appeared discolored and stained. Had he been out in the rain in that pair recently? he wondered to himself. He turned on the closet light for a better look. It was then that he noticed that the discoloration extended beyond the one pair. Several pairs were stained and speckled. As he picked up one of the shoes for a closer inspection, his nose quickly informed him that rain was not to blame here. This unmistakable odor could only be attributed to one thing: *cat pee.*

"Sharon!" Steve called out.

Steve and Sharon now had a problem that needed to be addressed right away. They loved their cat and were willing to be patient as he readjusted to life in their home, but accepting his spraying was asking too much. Steve phoned the vet while Sharon cleaned the kitchen. Wisely, Henry chose to remain out of sight.

The vet wanted Henry to be brought in for a urinanalysis. Instead, Sharon chose to collect some of the urine that had puddled on the stove so that she wouldn't have to deal with Henry for the time being. She was much too frustrated with him. Steve agreed to drop the container of urine off at the vet on his way to work. Meanwhile, Sharon tried not to allow herself to get angry with Henry. After all, there was a slight chance that the spraying was medically related. It was extremely doubtful, because Sharon knew that cats with bladder infections don't spray vertically, they urinate horizontally; but she was willing to give him the benefit of the doubt (at least until the test results came back).

The news from the vet was just as Sharon had imagined—Henry didn't have a medical problem. The vet recommended that Sharon call me.

I sat in Sharon and Steve's living room, listening to their story of Henry's unfortunate adventure. The cat had yet to make an appearance.

"I love Henry, but I can't live like this," Sharon said as she shook her head and glanced toward the kitchen.

"Being lost for such a long time could cause him to feel the need to reaffirm his territory," I offered. "Don't give up on him yet, because we have many options open to us."

"Thank goodness," Steve replied, and then took up Henry's story where Sharon had left off. As I listened, the overwhelming odor of cat urine kept finding its way to my nose. It was so strong, in fact, that for a moment I wondered if Henry had sprayed the very chair I was sitting in. The odor was so intense that it smelled like—

Just then Sharon tapped me on the knee, interrupting my thoughts.

"Here's Henry now," she said as she pointed to the black cat tiptoeing into the room.

The cat approached and jumped onto the coffee table. With his back toward me, he stood facing Sharon and Steve.

"Henry, come here on Mommy's lap," Sharon said, lightly tapping her thighs with her hands. There was no response on Henry's part. He continued to stand in the middle of the coffee table. Sharon continued. "What's my kitty up to?" she asked, and the cat then pricked his ears toward her. "Henry, this lady's here because you've been bad." Henry didn't move. "That's right, a bad kitty."

Just as Sharon said "bad kitty," Henry raised his tail up and flicked the tip of it a couple of times in response to her. As his tail went up, I noticed something that didn't make sense. Well, to be more precise, I noticed *two* things. No, it couldn't be, I thought. I quickly glanced back over my notes, which included a copy of Henry's medical records. What I saw on the record didn't match what I was seeing on the cat: *testicles.* Henry had testicles! Yet his medical record stated that Henry was neutered. But I was definitely staring at a pair of unaltered testicles. I reached across the table and petted Henry, and then tried to unobtrusively feel them for myself, just to make sure that my eyes weren't playing tricks on me. Yep, they were testicles, all right. Henry quickly glanced back at me, obviously surprised at the liberty I'd just taken.

I cleared my throat and looked up at Sharon and Steve. "Henry's medical records indicate that he was neutered," I said.

"Yeah," Steve replied. "Oh, we did that a while ago. He was about six months old."

Okay, here goes, I thought. There was no way to ease into this. "This is an intact cat," I said.

"What do you mean?" Sharon asked. They both look very puzzled.

"This cat," I continued, while pointing to the cat's back end, "has testicles."

"He can't," Steve stated.

"See for yourself," I answered.

"What the . . ." Steve began as he turned the cat around. "He does!"

I took a deep breath and leaned toward Steve and Sharon. "This isn't your cat," I said with as much sensitivity as I could.

"What?" they both replied in unison.

Sharon shook her head as if to try to shake the confusion away. "Is there ever a case where a cat's testicles ever grew back after . . ."

"Sharon, don't be ridiculous," Steve interrupted.

"This certainly explains the spraying behavior," I said. "Besides, the urine smelled so strong to me, so much like tomcat urine. Now I know why."

The cat, probably tired of having his private parts being the topic of conversation and inspection, jumped off the coffee table and started walking out of the room.

"There's something else," I said. "Before I saw that he was intact, I just happened to noticed that he didn't respond at all to his name, and of course, now we know why. But I did, however, notice that he responded to a different name. Watch this," I said, and turned toward the retreating cat. "Henry?" I called out. No response. "*Kitty?*" This time the cat stopped and turned toward me.

Sharon, who had moved to the edge of the couch when I'd told her that this wasn't her cat, now sat back and blew out a breath of air. "What does that prove?" she asked.

"I think this cat's name, or at least nickname, is *Kitty*. Not terribly original, but perhaps he's an outdoor cat that someone routinely fed and called him Kitty just for lack of a better name." I continued, "When you were speaking to him, Sharon, I noticed that he never responded when you said, *Henry*, but there was always a reaction to *Kitty*."

Steve nodded his head in agreement, and then added, "This is making sense now. This cat never acted like Henry at all but we just kept attributing it to what he'd been through."

There was a sudden sniffling sound. Steve and I both looked over at Sharon. Her eyes were filled with tears. "I thought that was Henry," she said through increasing sobs. "I was so happy to have him back. This means that Henry's still lost, maybe even dead. How could I have thought this was Henry? How could I have not known my . . ." Sharon's words became inaudible as she put her face in her hands and cried.

Steve got up from the couch and disappeared into the kitchen. He returned seconds later with a tissue, which he handed to his wife. And though he was trying to hide it, I heard a couple of sniffles from him as well.

I asked to see some pictures of Henry. Sharon slowly stood up and walked in an almost zombielike fashion to the entertainment center across the room. There, she bent down and pulled a small photo album off one of the lower shelves. The reason I wanted to do this was to see for myself how similar the cats looked. I was hoping that the cats were spitting images of each other (at least from the front) in order to help Sharon and Steve feel a little less embarrassed about misidentifying this cat.

As we flipped through the pages, it was quite amazing how similar the two cats' faces and body types were. I think Sharon and Steve wanted their cat back so desperately that they found ways to explain away any inconsistencies.

The problem the Ruhls now faced was what to do about this cat. Even though the cat obviously wasn't Henry, they were attached to him. Steve took his wife's hand and said that if she wanted to keep the cat, it was all right with him.

"You really should check with the humane shelter, though, just in case he belongs to somebody," I said. "And, he definitely needs to be neutered immediately."

They both nodded in agreement.

"I am curious about one thing, though," I said. "When you took this cat to the vet, how come nobody noticed that he had testicles?"

Sharon thought for a moment, and then answered that their regular vet had been out of town and someone unfamiliar was substituting for him. "I guess he just overlooked it," she surmised. Quite an oversight!

Before leaving, I spent time going over a behavior modification plan for the Ruhls to begin in regard to the spraying problem of this cat. Neutering was, of course, first on the list. The Ruhls also had to begin getting to know this cat as an individual, and not a Henry clone. They had to learn his likes and dislikes, his comfort level in terms of physical contact had to be respected, and they needed to know his playtime habits. He had a separate personality from Henry to which the Ruhls hadn't paid attention. They now had to get to know this little guy and begin training him.

As sad as Sharon and Steve were over the fact that their Henry was still missing, they both felt up to the task of welcoming this soon-to-be renamed cat into their lives.

I left, feeling both sad and elated. I was sad that the Ruhls had to suffer the heartbreak of losing their Henry a second time, but was thrilled that they had enough love in their hearts to keep this cat.

A week went by and I hadn't heard from the Ruhls, so I assumed all was going well. One month after my visit, though, I received a call from Sharon.

"Henry's back!" she said excitedly.

"What do you mean?" I asked, very confused.

Sharon went on to explain that everything had been going along just fine with the new cat (whom they named Hank). They'd had him vaccinated and neutered and they were following my behavior modification plan. Hank had settled in and was becoming quite comfortable. "Then one day," Sharon said, "I was at my neighbor's house. She lives about six houses down. She had moved here about three months ago. We were sitting in her kitchen having coffee. I looked out the window and saw a black cat sitting on the back porch. I asked my neighbor if that was her cat, and she said no. She told me that it was just some stray that began hanging around a few days after she moved in. My neighbor said she put food out for him every morning and let him sleep in the garage, but because of her allergies, she didn't want the cat inside."

Sharon said that her heart began racing in her chest as she stood up and walked toward the back door to get a closer look. "After the last experience, I was afraid to trust my judgment," she admitted. But when she peered through the window of the back door, the cat peering back at her was truly Henry. Sharon knew it in her soul this time. They locked eyes and Sharon quietly opened the door. Bending down, she called the cat's name. No sooner had the word "Henry" left her lips than the black cat started meowing loudly and crawled right up into her outstretched arms.

Sharon, while clutching the meowing cat in her arms, used her neighbor's phone to call Steve. Of course, when she told him that Henry was back, he didn't believe her.

"Does this cat have testicles?" he asked sarcastically.

"It's Henry! It's Henry! I don't know how it happened, but there's no mistake about it," she replied, though Steve made her check the cat's backside anyway. "No testicles!" she proudly declared. Her neighbor, who was sitting at the kitchen table watching this emotional scene, started laughing.

Henry and Hank now happily share a home together. Hank had very little trouble accepting the inclusion of Henry into his life. And Henry, so relieved to be reunited with Steve and Sharon, didn't mind Hank's presence just as long as Sharon's lap remained exclusively available for him alone. Hank seemed to understand the importance of that and took up residence in Steve's lap every evening while the family watched TV.

Every window in the Ruhl home has been fitted with sturdy new screens.

Freddie Knows Best

"Hello?"

"Is this Pam Johnson-Bennett? The one who works with disturbed cats?"

"Yes, it is."

"Well, you *must* help me," the man on the other end of the phone said emphatically. "My cat is terrorizing my wife. He's driving her crazy."

John Schreiber was forty-eight years old, vice president of an entertainment-related company, and newly married. His cat, Freddie—a seven-year-old white male whom John had rescured from the roadside years ago—strongly objected to the new Mrs. Schreiber, although for the first two months of the marriage, all had gone well. Freddie actually seemed to enjoy having Donna in his life.

"He'd curl up in her lap at night," John told me. "They really were happy together. Then suddenly, about a month ago, he started hissing at her. And then he began spraying her things. It's as if he's trying to drive her out of the house."

An appointment was scheduled for the following Saturday. In the meantime, John agreed to take Freddie to the vet for an exam.

The day before our appointment, John's vet phoned me to say he couldn't find anything medically wrong with Freddie, and he wished me luck in solving the problem.

When I arrived at the house, I found John sitting on his front porch, a cigarette wedged in the corner of his mouth. The cloud of smoke that drifted above his head stayed trapped under the porch ceiling. When I stepped out of my car, he rose and came over to greet me. The smell of tobacco was almost overwhelming. He had a very troubled look on his face. I shut my car door and we started walking toward the house.

"We had a bad night," John said as he bent down to put his cigarette out in the ashtray that sat perilously on the railing of the porch.

"What do you mean?" I asked, noticing the mound of cigarette butts piled up in the tray. It was obvious John had been sitting out here for quite a while.

John opened the front door and motioned for me to enter. He half-smiled, as if pretending I was there for a social visit.

We sat on the sofa. It took me a minute to place all of my equipment on the carpet near my feet and to get my notebook out. All the while, John watched me with a very concerned look. He stretched his arm across the back of the sofa and nervously tapped his fingers. I hurried to find my pen.

As I looked up, I noticed the chair across from me was completely covered in plastic. John told me that the only furniture in the living room Freddie targeted was that one chair. "It's Donna's chair," he said.

"Will Donna be here for the session?" I'd told John over the phone how important it was to have both of them present.

"Yeah, she's here somewhere. Excuse me a minute." He stood up and walked out of the room. I heard him calling for his wife. She responded rather curtly that she'd be right out. John walked back in the room and took his place on the sofa. "This whole thing has upset Donna so much. You're our last shot," he said sadly. "If Freddie doesn't stop spraying, then I'll have to find him another home. I just don't understand why he's doing this *now*." He reached for his cigarette pack on the coffee table, saw that it was empty, and tossed it back down.

Donna walked in and came over to greet me. Her handshake was quick and businesslike. She was much younger than I expected; she looked about twenty-five years old. Her back was straight and her hands rested in her lap as she sat across from her husband on an ottoman. She gave me a hard look—it felt like the meter was running.

I asked John to update me. He began the story of how wonderfully Freddie and Donna had gotten along initially. Then suddenly, a month ago, the cat sprayed her sweater. Donna had taken it off that night after work and draped it over a chair in the bedroom. The next morning, as she was folding the sweater to put it away, she noticed the urine odor. Two days later, Freddie sprayed one of her dresses. It was early in the evening; Donna had come home from work and left the dress on the bed while she showered before dinner.

"Freddie only sprays Donna's things," John said, giving his wife a sympathetic look. "He doesn't bother my clothes at all." Donna remained silent as John continued the story. "And he started spraying *that*." He indicated the plastic-covered chair.

"Donna, what about his behavior toward you directly?" I asked.

This must have caught her off guard—she appeared startled at the sound of her name.

"What?"

I repeated my question.

Donna shook her head and waved her hand in a gesture of dismissal. "He just hates me, that's all. And he's ruining my things."

John went on to explain how Freddie hissed a great deal at Donna whenever she first came home from work. He then followed her around while she undressed. That's when he would usually spray.

"So Freddie only hisses when you first come home from work?" I asked Donna.

"I'm not sure," she answered, and John looked at her, confused.

"You know he does. We've talked about how strange it is," John said to her. Turning to me, he explained, "He calms down later in the evening."

"Tell me about the chair," I said, pointing to the furniture covered in plastic.

John told me that Donna always threw her briefcase and coat on the chair when she came in the door. Then she'd sit on the arm of the chair, looking through her mail. "That's her routine," he said.

After we talked at length, I spent a half-hour with Freddie. He seemed like a playful, alert, and friendly cat. His love for John was obvious, as was his distrust of Donna.

I figured Freddie must be reacting to a scent on Donna's clothes. Cats can have strong reactions to certain odors. This required further investigating. I mentioned the possibility to John and Donna. John was intrigued. He said it made sense, because Freddie never bothered with any of Donna's other things, even if they were left on the floor—something she did very often, he added. Freddie's fascination was only with Donna's office wardrobe.

Donna thought the whole thing was nonsense.

"Do you work around any chemicals?" I asked her, trying to make some connection.

"No," she answered with a sly smile. I got the feeling she wanted to prove me wrong.

"Do you have a perfume you wear only to work?"

Again she said no.

"Your clothes always smell of Wayne's cologne," John said suddenly.

I watched Donna shoot John a disapproving look. "That's because I *work* with the guy," she answered impatiently. "He's my boss. What do you expect?" She turned to me. "I work in an advertising agency and I'm in meetings with my boss every day."

John leaned toward Donna. "But you started working for him a *month* ago."

"So?" she responded coolly.

"Well, actually, there is an ingredient in certain colognes that resembles the scent of a male cat's pheromone," I said.

"What?" John made a face.

I nodded. "It's in the musk-based colognes."

"His cologne does smell musky, and it's very strong," John confirmed.

Donna leaned forward, resting her arms on her knees. "This is silly," she said. "John, don't tell me you're actually buying this garbage."

"No, it makes sense." John patted her arm. "Think about it. I've told you so many times how I can smell his cologne. And it's only on the clothes you wear to the office."

"Well, what do you expect me to do?" Donna looked at both of us. "Just tell my boss to stop wearing cologne because it makes the cat pee on my clothes?" She started laughing. "Lady, you're as crazy as that stupid cat."

I didn't say anything. I've been called crazy before. I'm quite used to it.

"Donna!" John exclaimed. He looked wounded. Probably not because his wife just insulted a guest, but because she had called his cat stupid.

"Well, she is," she answered. "This is a waste of my time."

At the risk of making her even angrier, I offered a suggestion. "Donna, would you be willing to at least try an experiment?"

"Like what?" she huffed.

I wasn't hopeful, but I pressed on anyway. "Could you keep some clothes in a bag at work and then change into them at the end of the day before you come home? And put your work clothes directly into a garment bag or into the laundry?" I waited for her reaction, sure I had crossed the line again.

"Please try it," John pleaded.

Donna stared at John before finally throwing her hands up in defeat. "Okay, I'll do it. If it means the cat will stop ruining my clothes, it's worth a try." She glared at me and pointed her finger. "You better know what you're talking about, or I want our money back."

Before leaving, I gave Donna some suggestions for play sessions with Freddie to help win his trust again. This idea didn't go over all that well, but John promised he'd get Donna to participate in playtime with the cat.

Donna reluctantly participated in the experiment for three weeks, and, indeed, it did successfully stop Freddie from spraying her clothes. He showed no interest at all in the clothes she'd changed into right before coming home. Everyone seemed happy, even Donna.

John told me that Freddie was enjoying his daily play sessions. It seemed to be the only time he liked Donna. But he didn't quite trust her yet. I told John that three weeks was probably too soon to expect Freddie to completely forget the past. The fact that they were playing together was a good sign. And Freddie's absence of spraying was definitely an improvement.

Another two weeks went by and I received a call from John. His voice sounded as if he was working hard to control his emotions. He had some news he felt I should know. He and Donna were separating.

"I'm sorry to hear that," I said.

"I just wanted you to know that Freddie was right all along. He sure had Donna's number. You know how you told us about the pheromone in cologne?"

"Yes," I answered.

"Well, after Donna started changing her clothes before coming home, everything was great. Freddie didn't spray. Then, last Saturday, Donna said she was going out to lunch with Lexene, one of her girlfriends." John paused for a minute. "But while she was supposedly at lunch with her friend, Lexene and another girl stopped by the house, looking for Donna. I didn't say anything to Lexene. I just waited for Donna to come home."

"John, I don't think I'm the person you should be telling this . . ." I started to say.

"Wait," he interrupted. "It gets better."

Great.

"So Donna came home, and I asked her how lunch with Lexene was. She said it was fine. Even told me that Lexene said to say hello to me. Then she went to the bedroom, saying she was hot and wanted to take a quick shower. I was standing in the living room, just trying to keep my temper. I didn't know what to do. Then I saw Freddie come out of the bedroom, looking rather sheepish. So I walked in there and saw Donna's clothes lying in a pile on the floor, as usual. I picked up her shirt, and it was wet. Freddie had sprayed it."

John stopped again. He was either trying to control his temper or keep himself from crying. "Anyway . . . when Donna came back into the bedroom, I asked her where she'd really been. And then I told her Lexene had stopped by. She wouldn't admit to anything, even when I showed her the wet shirt. I never wanted to believe it, but now I knew for sure that she was seeing Wayne on the sly. I should have listened to Freddie sooner. She may have fooled me, but she couldn't fool Freddie."

So with the help of his cat, John discovered that his new wife had been cheating on him for almost two months. They separated, and John is now suing for divorce.

"Freddie's something," John said quietly that day on the phone, almost as if he were speaking to himself. "He knows women better than I do." Then, right before hanging up, he added, "Your information about the cologne made all the difference. I never would have been able to get to the truth without you. Thanks, Pam."

You're welcome . . . I guess.

○●○

Bald Sedona

Sedona was almost totally bald. All the eight-year-old Tonkinese had left was the hair on his head. The rest of his body hosted sparse, broken-off traces of patchy wisps. What had begun as a small bald spot on his stomach a little over five months before had now blossomed into a serious problem: the cat was now basically naked.

Upon noticing that first small spot, Sedona's owner, Dave Murphy, had brought him to the veterinarian immediately. When no cause could be found for the tiny hairless patch, a bitter-tasting cream was prescribed to prevent Sedona from continuing to lick the area.

A week later, Dave and Sedona were back at the vet because a second bald spot had appeared on the inside of the cat's right thigh. This spot was much bigger, though. Another exam was performed. To be on the safe side, blood tests were also done. At Sedona's age, doing a complete blood work-up was a good idea, in order to check on his kidney and liver functions as well as his serum protein levels. A CBC was also done to check the cat's blood cell count.

While the vet was drawing blood from Sedona, he mentioned to Dave that the bald spots might be caused by something other than a physiological reason. He mentioned the term *psychogenic alopecia*, a condition caused by obsessive licking.

"Psycho what?" Dave asked. "Does that mean my cat's got some kind of psychological disorder?"

"He could be feeling stressed or bored. I see this in cats who spend long periods of time in boarding kennels or who live in chaotic households."

Dave said that, except for a few new girlfriends in his life now and then, nothing major had changed in Sedona's environment. He couldn't imagine what in the world might be bothering his cat.

The vet said that the results of the blood tests would be in the next day; perhaps they'd find an answer there. In the meantime, to prevent potential intestinal blockage from all the hair Sedona was ingesting, he instructed Dave to regularly administer an oral feline hair-ball-prevention gel.

When the test results came back revealing Sedona to be the picture of feline health, the vet recommended that the cat be placed on a prescription drug shown to be effective in many cases of psychogenic alopecia. Dave agreed to try it, but after two months of drug therapy, Sedona's baldness continued to worsen. Different drugs were then tried, but nothing stopped the hair loss.

Out of desperation, an Elizabethan collar was put on the cat. This is a cone-shaped plastic collar that prevents an animal from reaching any part of his body with his mouth. The collar was somewhat successful in that it prevented Sedona from licking, but it was obvious he was miserable. Dave finally removed the collar because he could no longer tolerate watching poor Sedona trying to navigate around the house. In a short time, all the hair that had started to grow back was licked off again. There seemed to be no solution.

After drug therapy, Elizabethan collars, bitter-tasting creams, and pet sweaters, Dave took Sedona to another vet for a second opinion. The diagnosis of psychogenic alopecia was confirmed, only this time the vet recommended that Dave contact me to help Sedona stop his compulsive grooming.

Though he wasn't feeling especially optimistic, Dave figured he had nothing to lose, so he gave me a call.

Dave explained to me that Sedona had been a surprise gift from a former girlfriend, after Dave had lost a beloved cat to cancer. Sedona, a thirteen-week-old kitten at the time, made himself quite comfortable in his new home and was a great comfort to Dave during this time of sadness.

Being a very people-oriented cat, Sedona enjoyed racing around the house, sleeping on a big, overstuffed chair that Dave had found at a yard sale, and hanging out with his owner. When Dave's friends came over to watch football on TV, there was Sedona, in between the bowls of potato chips and pretzels.

"Sedona's always been just like one of the guys," Dave said during our initial phone conversation. "He's never been one of those prissy little cats who hides from everything. He loves everybody."

We scheduled an appointment for the following weekend, even though I was repeatedly informed by Dave that he was very skeptical about my being able to solve Sedona's problem.

Dave and Sedona lived on a very quiet street in the house that Dave had grown up in. He'd bought the place from his parents when they retired and moved to Florida.

I walked up the driveway and caught a glimpse of a curtain moving in one of the upstairs windows. As I got closer to the house, I saw another curtain moving in a different window, then a little feline head peered out at me. "Hi, Sedona," I called up to the cat.

Dave answered the door and welcomed me inside. He was a round man, with a big smile and dark, expressive brown eyes. "Here, let me get that for you," he said as he reached for my briefcase and bundle of cat toys. Sedona was right next to him, looking rather . . . uh . . . bald. He immediately came over to me, tail straight up in a friendly greeting.

As I glanced around at the house, I realized that Sedona must have been given that name in keeping with the entire theme of the house: the decor was southwestern.

Once we were seated in the den, Dave immediately started filling me in, while Sedona busied himself with investigating all the things I'd brought. First, he stuck his head in my open briefcase. Not finding anything of interest in there, he moved on to the interactive toys that lay near my chair. He soon had them in the center of the room and was in the process of dragging one off when I reached over and grasped the end of the toy just in time to prevent it from disappearing around the corner.

The history Dave provided certainly gave me the impression that Sedona was a happy cat, living a relatively carefree life with a loving owner. Based on Sedona's alopecia, though, it was obvious that this wasn't the case. Still, there were no other indications that the cat was having a tough time, so I temporarily ruled out stress.

I began to suspect that his problem stemmed from boredom.

"Have you ever seen Sedona in the act of this compulsive self-grooming?" I asked.

"No, never," Dave replied.

"Is he alone for long periods of time?"

Dave shrugged his shoulders. "Well, not really. I mean, just while I'm at work. I'm gone about nine or ten hours a day. Then, maybe three nights a week I'll go out, but it's not anything new. It's been like this since I first got him."

"When you're home, Sedona's always with you, right?"

"Always. Like a shadow," Dave answered, flashing that big smile.

"If you haven't ever seen him grooming himself, then he's either doing it during the day, when you're at work, or at night. Do you happen to remember whether you've noticed more bald patches on him when you come home at night or when you wake up in the morning?"

"I'll notice that some of the spots are bigger when I'm playing with him at night. So he must do all this during the day. But why only in the last five months?"

I looked up from my notebook. "We'll figure it out."

Dave and I talked a little while longer and then I spent some time with Sedona. I found him to be a playful, inquisitive, sweet-natured cat. "I'll get to the bottom of this," I whispered to him as I stroked his hairless back. Turning to Dave, I asked for a tour of the house so I could get a total picture of Sedona's environment.

I walked through every room, requesting that Dave point out any of Sedona's favorite or least favorite spots.

"He really only has one favorite spot," Dave said, pointing toward the window.

I walked over to the large, comfortable-looking chair sitting near the window.

"Sedona likes that chair a lot. When I'd come home from work, I used to find cat hair all over the top of it. Of course, now," Dave said sadly, "there isn't much hair to be found anywhere. But he likes to look out that window."

I stood in front of the chair and looked out. The window was on the side of the house, near the back, facing the side of a neighbor's house. There was not much to look at, though, since the view was obstructed by a privacy fence.

I noticed that the fence looked relatively new.

"Does Sedona, in general, like to look out all the windows?" I asked.

"No, mainly this one. I think it's the chair he likes."

As I gazed out the window, I found myself wondering why a cat would be interested in staring at a fence. The top of the fence was much too high above the window for Sedona to see any birds who might land on it.

"May I go outside?" I asked Dave.

He looked puzzled. "Yeah, sure."

I walked out and around to the side of the house. Dave followed.

"What are you looking for?" Dave asked.

"How long has this fence been here?"

"I don't know, six months maybe."

"Do you know what's on the other side of the fence?"

Dave gave me a weird look. "Yeah, the house next door."

I pressed on. "Do you have a ladder?"

"A what?"

"A ladder. I just want to see what's on the other side. It'll only take a minute."

Dave disappeared around the back of the house. When he returned with the ladder I noticed that the friendly smile I'd been seeing for the last hour was no longer evident. After he opened the stepladder, I climbed up and peered over the top of the fence. There it was! I now knew why poor Sedona was overgrooming himself. He was frustrated and bored. He had lost his great source of entertainment and excitement: the neighbor's very active bird feeder. Sedona wasn't interested in staring at a fence. He was interested in staring at what *used* to be there before the fence was put up.

With this new information, Dave and I went back to the house so I could set up a plan. I told him to buy a bird feeder and set it up right outside the window for Sedona. He'd been missing that stimulation. Then I showed Dave how to do play sessions with Sedona each night. This would allow the cat to enjoy "capturing" birds. Since Sedona was a Tonkinese, one of the more athletic breeds of cat, the vigorous game would be perfect for him.

I also suggested that Dave consider installing a fish tank with a cat-proof lid for Sedona's entertainment. "That way, if there aren't any birds around one day, he can amuse himself watching the fish. But along with the installation of the fish tank and the bird feeder, you have to provide regular play sessions with Sedona so that he can act on his predatory skills."

As I was leaving Dave's house, he promised he would go out that very afternoon to buy all the necessary items.

Weekly follow-up calls from Dave revealed that the new bird feeder and fish tank had done the trick. Sedona had stopped the excessive grooming, and his hair was beginning to grow back.

Six months later I received some pictures of Sedona in the mail along with a note from Dave. One picture showed Dave and Sedona sitting on the sofa together, watching the fish tank. Another showed Sedona sitting on his favorite chair, watching the bird feeder. The cat looked happy and *hairy.* In his note, Dave mentioned that when his friends come over to watch TV, Sedona still sits between the potato chips and the pretzels—only now he watches his fish tank instead of the game.

The Walls Have Ears

It was 7:30 on Sunday morning, and I was still enjoying the foggy last stages of deep sleep when the telephone roused me.

"Hello," I answered, trying to sound more awake than I actually was. While it wasn't unusual for clients to call early in the morning, it always seemed to happen after a late night out. Catching up on sleep just never seemed to work for me.

"Dr. Kramer said I should call you about my cat. My name is Mrs. Wendy Hearn."

I doubted that the vet had told her to phone me at 7:30 A.M., though. "How can I help you?" I asked as I rubbed my eyes and looked over at my own two cats, who were anxiously awaiting breakfast. My dog's tail began thumping the floor in happy anticipation of her morning walk. While they are usually polite enough to let me sleep late in the mornings, my animals all know that once my eyes are open, it is open season for my attention. Eye contact is my downfall when it comes to my pets, and an early-morning call is just the kind of opportunity they wait for.

The weather report the night before said that today's heat would be record-breaking. Glancing out the bedroom window, I noticed that everything already looked wilted. The quiet hum of the air-conditioning provided a comforting sound, lulling me back toward sleep. I suddenly realized I wasn't listening to Mrs. Hearn and turned my attention back to the conversation.

"Ralphie is climbing the walls," she was saying.

"You mean he's hyperactive?"

"No, he *really* climbs the walls. He also throws himself at them. It's as if he's possessed."

Snatching my notebook and pen from the nightstand, I jotted down Mrs. Hearn's phone number. I told her I'd call her back in a few minutes after I'd had time to take my dog out and track down my appointment book. The previous night I'd had a client emergency with an aggressive cat and hadn't gotten home until 3:30 in the morning. At the moment, I wasn't at all sure where I'd deposited my appointment book, purse, keys, or even shoes.

When I called back, Mrs. Hearn explained to me that her cat, Ralphie, a five-year-old Russian Blue, had begun having fits and hurling himself into the walls. Initially, it had started with him scratching at the walls and then attempting to climb them. That soon escalated into head-first suicide runs. When Mrs. Hearn called her vet, he'd suggested that Ralphie needed a scratching post. Mrs. Hearn had purchased one, and Ralphie happily used it, but he still continued his attacks on the walls. The vet then suggested that Ralphie might be bored and in need of more stimulation, so Mrs. Hearn purchased toys. Ralphie wasn't interested.

"I brought him in to Dr. Kramer," Mrs. Hearn said, "but he couldn't find anything wrong with Ralphie. That's when he suggested I call you."

"Does the cat do this all the time?" I asked.

"Much of the time. It's like he's trying to kill himself. Do cats do that?"

"I wouldn't think so."

"When can you come over?"

I flipped through my appointment book. "How about Tuesday evening?"

"No, no," she answered. "I'd like you to come over now, this morning. I'm really worried."

I am generally available for clients seven days a week, but it just so happened that I had designated this as my official day off. I had a lot of personal errands to catch up on, and my house was in dire need of a cleaning. "Today is really . . ."

"Please," she interrupted. "I can't stand to watch him do this anymore."

"Sure. How about this afternoon at 2:30?"

"Can't you come over right now?" she asked urgently.

I truly felt that it could wait until afternoon, but I agreed to come over in an hour.

"Make it a little sooner. Do you mind?"

"Mrs. Hearn, I'll leave here as soon as I can, but you have to give me time to get there. You live thirty minutes away. I promise I'll hurry."

"I'll be waiting for you," she said and hung up the phone.

"So much for my day off," I said to my dog, who had returned to bed and was stretched out comfortably on my pillow. "I'll be leaving now," I said to her as I walked out of the bedroom. "No, don't bother getting up."

Mrs. Hearn was waiting at the door for me when I arrived. A petite woman, she looked to be in her early fifties. As I walked into the house I saw Ralphie sitting on the floor behind his owner.

"Thank you for coming," she said, ushering me in and quickly closing the door against the oppressive heat outside. "He scratched the walls again since I spoke to you. I think he's nuts!"

I looked over at the quiet cat, who hadn't made any attempt to move. Mrs. Hearn followed my glance. "I'm telling you, he was scratching the walls," she said defensively. "He's just being spiteful now to make me look bad."

I raised an eyebrow at her, then peered back at Ralphie. He gave me a slow wink, turned around, and walked away.

"I'd like to get a little more history on your cat," I said as I followed Mrs. Hearn into the living room. I sat on one of the overstuffed chairs while Ralphie chose the chair opposite me; he was ready to begin the session. Mrs. Hearn sat on the couch, carefully keeping her eyes on the cat. Her expression seemed to be that of a parent disappointed in her child's refusal to perform for company. Ralphie took it all in stride and was content to sit and watch the proceedings. He even managed a dramatic yawn.

"I don't understand it. He was acting up this morning. I don't know why he's being so calm now," Mrs. Hearn said. She shook her finger at Ralphie. "You're making me look foolish."

"I'm sure he's just distracted by my being here," I suggested.

"No, he's being spiteful."

For the time being, I decided to ignore her insistence that Ralphie was doing anything deliberately to her. "When did you first notice this behavior?" I asked.

"A week ago. He started staring at the walls. It was very strange. He'd just face the wall over there and look at it for several minutes." She pointed to the wall across from me. "Then he began pacing in front of it. The next day he was scratching at the wall as if he wanted to tear through it. Finally, he'd just charge right toward it."

Ralphie yawned again, stretched out, and then curled up in the chair for a little nap. Mrs. Hearn rolled her eyes in apparent frustration at her cat's lack of cooperation.

"Have there been any changes in Ralphie's life recently?"

She shook her head. "No, things have been pretty much the same. My husband died five years ago, and Ralphie's been such a comfort for me." She stared at her cat. "That is, until recently." Suddenly she rose to her feet, crossed the room, and scooped the sleeping cat up in her tiny arms. "I've had just about enough of you, young man," she scolded. "Now, show the lady what you've been doing."

"Mrs. Hearn, I'd prefer that you not force Ralphie. I'd like to give him time to get used to me."

She glared at the cat and then at me. "I know what I'm doing," she snapped, then proceeded to deposit her cat on the floor in front of the wall.

For a petite woman, she carried herself with a great deal of authority. Ralphie sat on the floor and gazed at me. I certainly sympathized with him.

"Let's allow Ralphie to relax," I suggested. "I'd like to try a play session with him so I can get to know his personality a little better."

Mrs. Hearn gave Ralphie one last look of disgust and then walked back to the couch. I positioned myself on the floor near the wall. Ralphie perked up immediately when I pulled out an interactive toy from my case. We began to play. I moved the toy along the wall and even ran it up the side, but Ralphie was only concerned with our game. At the moment, the wall itself held no interest for him.

As we played, I continued to get more information from Mrs. Hearn. Apparently, these head-on wall attacks occurred at all times of the day or night.

After our game, I asked Mrs. Hearn to show me the exact places that Ralphie attacked the wall. I wanted to see how close to the window they were, but the areas she pointed out were right in the middle of the wall, far from any windows or doors.

So far, during my visit, Ralphie appeared to be a friendly, alert, well-adjusted cat—although I wasn't doubting Mrs. Hearn's claim about his odd behavior.

As we were walking through the house, Mrs. Hearn stopped suddenly and began searching for something in her pocket. "Wait a moment, dear. I forgot my eyeglasses," she said, and she headed back to the living room.

Coincidentally, at that moment my pager went off. "Mrs. Hearn," I called out. "May I use your phone, please?"

"Help yourself," she answered. "It's in the kitchen."

I quickly walked into the kitchen, found the phone, and began dialing. Almost instantly, Mrs. Hearn appeared in the doorway, looking extremely excited. "He's doing it, he's doing it!" she shouted. "Hurry! He's going crazy."

I immediately replaced the receiver and followed her back into the living room. There, I saw Ralphie sitting quietly in the center of the room. He looked surprised to see us.

Mrs. Hearn raised up her arms. "He *was* doing it. I swear he was."

Ralphie sat motionless as his owner carried on about his being uncooperative and costing her "money for nothing."

I reassuringly put my hand on Mrs. Hearn's arm. "We'll figure this out. Why don't you go make yourself a cup of tea and relax. After I return this phone call, I'd like to spend some time with Ralphie by myself. Is that okay with you?"

"Sure," she said, not sounding very hopeful.

During my private session with Ralphie, he continued to appear to be a well-adjusted, even-tempered cat, but I was determined to find the solution to the problem.

I walked into the kitchen where Mrs. Hearn was finishing a cup of coffee. "Could we take a tour of the other rooms that Ralphie uses?"

Mrs. Hearn nodded, placed her empty coffee cup in the sink, and mumbled something barely audible. I did manage to catch "miserable cat . . . can't count on him for anything . . . making me look like a crazy old lady." I silently followed her out of the kitchen.

I toured the first and second floors, carefully checking out all the areas where Ralphie had connected with the walls. After spending time with the cat and hearing the description of his behavior, I had a hunch about what might be going on. I asked to tour the basement.

"The *basement?*" Mrs. Hearn asked incredulously. "Why on earth do you need to go down there? Ralphie isn't allowed in the basement. Are you here to stop his behavior or not?" She made no attempt to hide her displeasure.

"Of course I am, and I have a hunch . . ."

"A hunch!" she interrupted. "Is that all you can offer? Well, don't waste your time. There's no reason to go downstairs." The frown lines on her forehead deepened. Even though I wanted to be a bit more certain before mentioning my suspicion, I knew if I didn't give her an answer soon, she'd be showing me to the front door.

"The reason I'd like to check the basement is to look for any signs of mice."

A look of horror crossed her face, but I'd been expecting that.

"Mice?" she shrieked.

"Based on Ralphie's behavior, I'm wondering if there isn't a mouse in the wall. I think Ralphie might be hearing the noise and trying to get at it through the wall."

"That's ridiculous. Why don't you just admit that you have no idea why he's behaving this way?"

"Mrs. Hearn," I said very politely, "it's because of Ralphie's behavior that I suspect this. It seems logical to me, and I'd like to rule this possibility out first. If I don't see any signs of mice, then we can explore other options."

She glared at me, and then let out a heavy sigh. "I've *never* had mice. I keep a very clean home." She abruptly turned to leave the room, motioning me to follow. "We'll go look in the basement, just to prove that you don't know what you're talking about."

What a Sunday this was turning out to be. "Thank you," I meekly responded. "Do you have a flashlight?"

"No," she answered impatiently.

"I'll get one from my car."

I was retrieving my flashlight from the glove compartment when

Mrs. Hearn appeared in the doorway. "Hurry, hurry, he's doing it again! Hurry, Pam!"

I ran up the driveway and into the house. "Where is he?"

Mrs. Hearn pointed toward the dining room. "In there."

As I silently peeked around the corner into the dining room, I saw Ralphie curled up on one of the chairs, casually licking his left front paw. I felt Mrs. Hearn's presence behind me and turned around to get her reaction.

"Why is he doing this to me?" she asked. "He was going crazy a few seconds ago."

I looked back at Ralphie. *I will get to the bottom of this,* I thought as I watched him luxuriously grooming himself. "Let's go to the basement," I said to Mrs. Hearn, and I tightened my grip on the flashlight. My hunch had better be right.

At the top of the basement stairs, Mrs. Hearn flipped on the light, and we began to descend. "Even the cleanest homes can have mice," I assured her, but either she didn't hear or she chose to ignore me. Perhaps that was best.

At the bottom of the stairway, Mrs. Hearn searched for the switch to light the rest of the basement. The light it cast was very dim, causing the large room to appear quite sinister and full of shadows. Mrs. Hearn must have noticed this also because she gave me an uncertain look.

"Are there any other lights down here?" I asked.

"Not that I know of. My husband was the only one who ever came down here. This is where he kept his tools."

Her answer gave me more reason to suspect I was on the right track. No one had been down here to notice any signs of infestation; mice would have a free reign.

Armed with my flashlight, I slowly walked around the basement. The light hit a maze of pipes, various dusty old boxes, some broken chairs, and a few leftover tools. Old rags hung from bent nails on Peg-Boards, and messy paint cans were piled in corners, blanketed by years of cobwebs. My light caught a surprised spider as it scurried into one of the holes in the Peg-Board. I felt a little shiver. I guess I must have watched too many scary movies as a child, because basements never fail to give me the creeps. Neither do spiders.

Nevertheless, I continued on my search, combing every inch of the basement with my flashlight and silently praying I wouldn't be disturbing any more spiders. Suddenly the light came across exactly what I'd hope to find. In an ancient, rusty sink lay the remains of an old bar of soap, looking very chewed. Surrounding the dried-up soap were a substantial amount of scattered fecal droppings.

Mrs. Hearn, who was right behind me, gasped when she saw the sight. "What is that?"

"I believe a mouse has been nibbling on this soap," I said as I pointed to the shriveled green lump. "And these, I'm afraid, are his fecal droppings."

"That's impossible," she said. "I told you I don't have mice."

In spite of overwhelming evidence, I replied, "Why don't we continue to look for them?"

"*Them?*" she asked in horror. "You mean there's more than one?"

"Based on all the rooms you've shown me where Ralphie has attacked the walls, I'd say there has to be more than one."

"Oh, no," she said. "This is unthinkable."

As I moved on through the basement I found several areas that indicated the presence of mice. It was indeed looking like a serious infestation.

After being convinced that she did indeed have mice, Mrs. Hearn led the way back upstairs. Ralphie met us at the top of the stairs, rubbing against our legs as we reentered the kitchen.

"So," said Mrs. Hearn reluctantly, "I guess maybe my Ralphie really doesn't have a behavior problem?" She bent over and scratched the affectionate feline behind the ears.

"I think he's just very frustrated, Mrs. Hearn, because he's hearing all this movement behind the walls and he can't get to the source. Cats are hunters, and very often the first indication of nearby prey is *sound.*"

At Mrs. Hearn's request, I spent the next half-hour phoning exterminators. For some unexplainable reason, luck was finally on my side, and after several unsuccessful calls, I was able to talk to an actual person instead of an answering machine. His office wasn't officially open on Sundays, but he just happened to be there doing some paperwork. He agreed to come over first thing Monday morning. At this point, Mrs. Hearn got on the phone and tried to insist that he come over right away, but the exterminator was not going for it. Eventually, Mrs. Hearn let the poor man off the phone and turned her attention to my final instructions.

I gave Mrs. Hearn some play-therapy techniques to use with Ralphie to help distract him away from the walls. He needed to have some successful captures to ease his frustration.

As I said good-bye to Mrs. Hearn, she apologized for doubting me. "I still can't understand how there could be mice in his house," she said.

"I'm just glad we found the cause of the problem," I answered cheerfully. "Soon Ralphie will be a much happier cat."

As I headed toward the door, I thought I caught a glimpse of something small and gray scurrying around the corner. I looked at Mrs. Hearn to see if she had noticed, but she was too busy unlocking the front door and commenting about the oppressive heat. Ralphie, on the other hand, caught my eye, and we exchanged knowing glances.

I decided not to mention anything to Mrs. Hearn. I figured it would just upset her further. The exterminator would be there soon, and he'd do a thorough investigation of the entire house.

"Bye, Ralphie," I said, bending down to pet him. "You are quite possibly the most determined mouser I've ever met."

○●○

The Cat Burglar

Sabrina had once been a gentle giant. At least, that's how her owners used to describe her. But these days she was skittish and timid.

I was called to the home of Patty and Joe McGill at the request of their veterinarian, Dr. Kreitler. Sabrina, a formerly very healthy cat, was now losing weight, despite the fact that she had a normal appetite. She had also experienced a personality change in the past month. Once a friendly and playful cat, she was now just a shadow who darted around corners and slept in closets.

When the weight loss was first noticed, Dr. Kreitler performed blood tests on Sabrina. All results came back well within the normal range. The vet then switched her to a prescription food and also sent home a vitamin supplement. He questioned the McGills about any possible changes in their home that might be upsetting the cat, but there were none worth mentioning.

When Sabrina continued to lose weight, even with the calorie-dense prescription food, the McGills brought her back to the vet for further tests and radiographs. Again, the results came back normal.

Dr. Kreitler called me to discuss Sabrina's case. He said that the McGills were scheduled for an appointment at one of the veterinary university hospitals, but that he would first like me to pay a visit to the home. Before subjecting poor Sabrina to another battery of tests, he wanted to be certain there was no behavioral connection. It was a long shot, but he felt there was nothing to lose. All avenues had to be explored. Dr. Kreitler was so concerned about finding the cause of Sabrina's problem that he planned on paying for my consultation himself. I offered to see Sabrina at no charge.

The day of my scheduled appointment was bitter cold. The weather forecast had predicted that temperatures wouldn't climb out of the single digits. It had rained that morning, and the streets were quickly becoming icy and treacherous. I truly didn't look forward to being on the road, but I knew how worried everyone was about Sabrina.

And so, braving the bone-chilling cold, slick streets, and whatever other crazy drivers were also risking travel, I made my way to the McGill home. It wasn't until I pulled into their driveway that I realized how tightly I'd been gripping the steering wheel. I released my iron grasp, shook my hands to revive the lost circulation, and then braced myself to face the blast of cold air once I opened the car door.

Usually, clients are anxiously awaiting my arrival and are at the door to greet me before I can even ring the bell. As I gingerly walked up the icy driveway, I was hoping the McGills would be no exception—I didn't want to stand outside one second longer than necessary. But, as luck would have it, the McGills must have been at the absolute opposite end of the house, because it took a good three minutes before they answered the door. It seemed like hours.

"You must be Pam," said Joe McGill as he grasped my numb fingers in a warm, friendly handshake.

"Yes," I answered through chattering teeth.

"Please come in," said Patty McGill. "It's freezing out there."

I whispered a prayer of thanks when I saw that our session was to be held in the McGills' cozy kitchen. We sat down at the table, where Patty had laid out cookies and muffins. The aroma of coffee brewing filled the room. I wrapped my frozen fingers around the warm mug Patty offered me.

"I've never had a pet therapist in my home before," Patty said as she pointed to the food. "I'm sorry these aren't homemade, but I don't cook at all. We spend very little time in the kitchen, actually. We're strictly a takeout and delivery family."

"Dr. Kreitler has a lot of faith in you," Joe said. "We're really hoping you'll be able to figure out what's wrong with Sabrina. Should I go get her?"

"No," I answered. "Let's talk first. I'd like to get more background information."

For the next twenty minutes, Patty and Joe filled me in on Sabrina's problem. I had already received a case review from Dr. Kreitler, but I needed to hear how the McGills were viewing the situation. Their version was almost identical to Dr. Kreitler's. What we had here was a cat with a total personality change and an unexplained weight loss. The one new and significant piece of information I learned from the McGills was that Sabrina was an indoor/outdoor cat. There was a pet door in the kitchen so that she could come and go as she pleased. I thought that perhaps she had suffered some trauma outside, physical or emotional, that had caused the problem. The McGills assured me she hadn't, but I explained to them that it could have happened without them being aware of it. This is especially true in the case of emotional trauma.

When it was time to meet Sabrina, Joe led the way, searching behind chairs and under tables. "In the old days, she would have been the first one to meet you at the door," he said sadly. He bent down to search under the sofa. "But now we hardly ever see her."

The McGills described Sabrina as having been an extremely sweet-tempered cat. She could always be found within arm's reach of either Patty or Joe. Her toy basket overflowed with just about every kind of cat toy on the market, and her owners regularly engaged her in interactive play.

There was no sign of Sabrina on the first floor, so we climbed the stairs to begin checking the bedrooms. As we walked, I remember thinking how warm and inviting the house appeared. And, as always, I also looked at the home from a cat's point of view. Some homes aren't very cat-friendly. They may be too sparsely decorated, offering no secure hiding places. Owners who won't allow the cat on furniture and fail to provide any cat trees or perches also create a not-very-cat-friendly environment. Cats need elevated places to view their territory from a "safe" location. The McGill house, on the other hand, seemed to have everything a cat could want. Patty informed me that Sabrina had always been allowed on any piece of furniture she desired. Several carpeted cat trees sat near windows, plush pet beds could be found on the tops of two tall wooden cabinets, and scratching posts seemed to be everywhere. This certainly appeared to be a cat-friendly home—and for Sabrina, it used to be. The question was, what had changed all that?

I had gone down the list of every possible change or upset I could think of, but the McGills just shook their heads "no" each time. As far as they knew, life had been going along normally.

The McGills seemed like very nice people. Although extremely concerned about their cat, they were calm and quiet. I always watch for parallel personality changes between cats and their owners. Sometimes owners who become stressful and nervous can cause their cat to react accordingly. Patty and Joe didn't display any behavior that would cause me concern.

"Well, she has to be in here," Joe said as we walked into the master bedroom. Two eyes were peeking over the top of the laundry basket that sat in the corner. "Well, hi there, Sabrina. We've been looking all over for you." He turned to me. "Do you want me to get her out?"

"No," I answered. "Let her stay where she feels more secure." I had brought one of my interactive toys upstairs with me. "Hopefully I'll be able to entice her with this." I sat down on the carpet near the basket. I looked up at Joe and Patty, who were anxiously hovering over me. "Why don't you two sit down. Let's just keep this whole thing very casual. I want Sabrina to feel comfortable having me in the room."

Joe and Patty obediently sat on the carpet. "Should we talk or keep quiet?" Joe asked.

"You can talk," I answered. Slowly, I moved the toy back and forth within Sabrina's field of vision. I didn't want to excite or overwhelm her, so I kept my movements subtle. Sneaking a peek at her, I saw that her eyes were following the toy with interest. She'd glance suspiciously over at me every now and then, but I stayed quiet, and within a few minutes, Sabrina crawled out of the basket and a little closer to the toy.

Even though it was obvious that she wanted to play, her fears wouldn't let her. She retreated back to the laundry basket. Just the fact that she showed some interest in the toy was a positive sign, though, and I was pleased.

As I observed the cat, I realized how different she looked from the many pictures of her displayed throughout the house. At one time she had been a puffy, charcoal-gray cat with a round, black face. Her glossy coat had been long and silky. The Sabrina who looked back at me from the safety of her hiding place was noticeably thinner; her eyes were sad and frightened, and her coat was dull and dry. She looked more like a cat who had been living a hard life in the streets than one enjoying this cozy home. After a few minutes, Sabrina reappeared and again began making attempts at capturing the toy.

After a good half-hour of playtime, I suggested we leave the room so I could take a tour of the house. "I'd especially like to see any areas Sabrina favors or fears."

We carefully went through the house, stopping long enough at each cat tree, bed, and other favorite spot, where I looked out windows and checked for anything that might provide a clue. I also checked the locations of the litter boxes; there was one on the main floor and one upstairs.

Back in the kitchen I opened the door where the pet door was located and looked around. The cold air hit hard. I closed the door, and went into the front hallway for my coat, and then headed outside. It was even colder than when I'd arrived, but I was determined to find a clue to Sabrina's behavior. I walked around the house, looking for any signs that might reveal the presence of another cat—or any other animal.

"Do your neighbors have dogs or cats?" I asked the McGills after I came back inside.

"The neighbor on the left has a dog, but he's a very old one," answered Patty. "The lady has to carry him outside because he can hardly walk."

Well, that ruled out my theory of a neighbor's big, menacing dog scaring Sabrina, although I was still considering the possibility that a strange dog might have gone after her. "Does Sabrina go outside as much now?" I asked.

Joe shook his head. "With this cold weather, she prefers to stay indoors."

"Was it this way last winter?" I asked, although I doubted it was the cold weather that was keeping Sabrina inside.

"It wasn't as cold last year," Patty answered.

"What time do you normally feed Sabrina?" I asked.

Joe looked up at the clock. "Any time between now and an hour from now."

The McGills used to leave dry food out all the time, then supplement it with canned-food meals in the morning and evening.

Sabrina now refused to eat the dry food. Patty said the cat would walk over to the bowl, growl at it, and then run away. They had tried several brands of food, but nothing made a difference; they never knew from one day to the next what Sabrina's reaction would be. Some days she'd take a few nibbles of the dry food; other days she'd fiercely growl at it, or she might just hiss as she approached the bowl, whack it with her paw, and then run off.

"She loves her canned food, though," Patty said. "The sound of the can opener brings her running into the kitchen, no matter what."

I asked Patty to feed her now so I could watch her behavior. I wasn't sure she'd eat with me in the house, so I sat very still at the table.

Patty opened a can of food, and sure enough, there was Sabrina. She charged into the kitchen, then stopped suddenly when she saw me. The aroma of food must have outweighed her apprehension, because she decided it was worth the risk of having me nearby. She ate hungrily and then bolted out of the room, presumably back upstairs to her hiding place inside the laundry basket.

"She never acted this way before," Joe commented. "It used to be that after she ate, she'd come into the bedroom to hang out with us. Now she just runs off to hide."

"Do you normally stay in the kitchen when she's eating?" I asked.

"No," answered Joe. "When we feed her in the morning, we go back upstairs to get ready for work. Then, at night, we feed her when we come in. While she's eating, we get changed and go through the mail."

I had a feeling that whatever was causing Sabrina to be so skittish had to be connected to a traumatic experience outside. I wanted to close down the pet door and have the McGills turn her into an exclusively indoor cat for a while. I also noticed how close to the pet door the dry food bowl was kept. Her canned-food meal was placed closer to the sink, away from the door. That may be why she growled only at the dry food.

I explained this to Patty and Joe. For now, I wanted Sabrina fed in the room where she felt the most secure. Patty said that was either the bedroom or the bathroom. I also wanted one of them to stay in the room with her when she ate, to give her a more secure feeling. In addition, I had behavior-modification exercises I wanted them to perform with Sabrina. They said they'd be willing to lock the pet door and move the food bowl. Their response to behavior-modification exercises was very positive. I was beginning to feel hopeful.

After demonstrating the specific therapy exercises, I made an appointment to return in four days to check on Sabrina.

As we were saying good-bye at the door, Joe happened to mention that they were going out of town for a wedding the next day and would have to board Sabrina overnight. "Do you think that will set her back?" he wondered.

I didn't feel that putting Sabrina in a foreign environment would be good for her at all, especially since they were going to board her at the animal hospital. This was a place where she'd recently endured so many unpleasant and scary procedures. "Is there anyone Sabrina trusts who could pet sit for you?" I asked.

"No. Most of our friends are afraid she'll get sick or starve to death while in their care," Joe responded with a weak smile. "They don't want to be responsible for her."

I really did not feel it would be good for Sabrina to be boarded. "Would you feel comfortable if I came and took care of her?" I asked. "You don't know me very well, so I'd understand if you'd prefer not to give a stranger access to your home. I do think it would be better for Sabrina to stay in her familiar surroundings, though."

Joe's face lit up. "Would you really do that?"

"Dr. Kreitler thinks the world of you," said Patty. "So of course we trust you."

"It would also give me time to work with her," I added.

Patty left the room and returned with a key. "This is for the kitchen door," she said as she handed it to me. "The front door lock is very tricky, and you might have trouble opening it. I'm sure you'll find the kitchen door is much easier."

Joe said they'd feed Sabrina her breakfast before leaving, so I'd only have to come at night for dinner and then breakfast the next morning. They were both so grateful, I was happy to do it.

The next day, Nashville was treated to snow; by late afternoon, there was almost an inch on the ground. My drive to the McGill house was slow. Nashville drivers aren't used to snow, so I kept my distance from the cars around me. When I finally arrived, I was ready to unwind with some play therapy—I figured it would do both Sabrina and me some good.

As I walked around the side of the house toward the kitchen, I noticed a few paw prints in the snow near the door. Were they Sabrina's? Had the McGills disregarded my recommendation that the pet door remain locked? I bent down and checked the door. The pet door was locked. I looked again at the paw prints before going inside.

In the kitchen, I opened some canned food for Sabrina's dinner. I also noticed that the McGills had moved the dry-food bowl closer to the sink, as I'd instructed. With a bowl of moist food in one hand and an interactive toy in the other, I went in search of Sabrina. I found her in the upstairs bathroom, hiding behind the hamper. Setting the bowl of food down, I backed out of the room and sat on a rug in the hallway. To help her get used to me, I casually chatted away about this and that, telling her all kinds of uninteresting facts about my life. And it certainly did seem to relax her. After some initial reluctance, she came out and ate her dinner. Then I was able to engage her in a play session.

After playing for about half an hour, I went in to check on her litter box, to see if it needed cleaning. As I passed her dry-food bowl I noticed it was empty. "I'll refill that for you," I said, and, scooping it up, I headed back to the kitchen.

I was washing out the food bowls in the kitchen sink when I

heard a thump at the kitchen door. It wasn't a knock; it was a soft thump. Then I heard it a second time. This time it was followed by a scratching sound. I peered through the curtains on the door. Looking down, I spotted a black cat with a white spot on his head and four white paws.

Very quietly, I walked to the other side of the kitchen and shut the swinging door that closes the kitchen off from the rest of the house. I tiptoed back to the pet door, silently removed the plastic plate that locks the flap, and then hid on the other side of the kitchen to see if I was about to have a visitor.

I didn't dare make a move. I suspected the cause of Sabrina's problem was on the other side of the pet door. The McGills had told me that they usually left the kitchen after they set out food for Sabrina; so they wouldn't necessarily know if a cat came in. "And Patty did say they don't cook, so they don't spend a lot of time in here," I said to myself and then realized I'd spoken out loud. I clapped my hand over my mouth, hoping I hadn't scared off the black cat.

Suddenly there was movement at the pet door. The flap moved inward, but no one came through. I waited. Then it moved again, and I caught sight of the white-socked paw touching down on the kitchen floor. I smiled.

Wishing I had a video camera, I watched the black cat sneak in the door and look around. He went over to where the bowl of dry food used to be. Disappointed, he turned toward me. "So you *have* been here before," I said to the little thief. "Sorry, but you've just been busted."

The black cat noticed the dry food over by the sink and began to walk closer. Then, thinking better of it, he stopped, looked at me again, and hissed. He must have known I was going to be the one to rat on him. He looked back at the food, hissed again, and then swiftly made his exit out the pet door.

I ran to the window and watched him run out of the yard. Normally, his black coat would have been the perfect cover for a thief in the night, but there was no hiding against the background of white snow.

After locking the pet door, I headed back upstairs to play with Sabrina again. I had good news for her.

Later that night, I called Dr. Kreitler at home and told him of my encounter with the four-footed thief. Sabrina could obviously pick up the intruder's scent on her bowl, and that's undoubtedly why she growled and hissed at the dry food. The black cat may even have been eating a good portion of her moist food as well.

"Why didn't Sabrina growl at the canned food if the black cat was eating that too?" Dr. Kreitler asked me.

"Because that bowl gets washed after every meal. I noticed that the dry-food bowl just gets refilled without being picked up and washed first."

Upon their return, the McGills were thrilled to learn that the cause of the problem had been revealed, but they weren't too pleased with the fact that a strange cat had been in their home.

Sabrina was retested for FeLV and FTLV, both highly contagious feline diseases. The results were negative.

Within a month, Sabrina was becoming her old self again. The pet door was sealed permanently, and she was beginning to gain some weight.

Joe and Patty tried to catch the black cat burglar, but he was too clever for them. Eventually, one of the rescuers with whom I work was able to capture him. The McGills knocked on all the doors in the neighborhood, but no one knew where the cat burglar lived.

We had him tested, vaccinated, and neutered. After some serious behavior work to help him over his cynical attitude problem, he made parole and was released into the custody of a loving family.

○●○

Romeo, Romeo

Rosemary Lockmiller woke up one morning to find a face staring at her through the bedroom window. Rosemary's bedroom was located on the second floor. The staring face belonged to a cat who sat precariously perched on a tree branch just outside the window. The cat, a dirty steel gray in color, was slender with a menacing face. He stared directly and seemingly unblinking at Rosemary as she sat up startled in her bed. She was so taken aback that she instinctively pulled the sheet up to cover her nightgown. After recovering from the initial shock, Rosemary regained her composure, got out of bed, and walked over to the window. She banged on the glass to chase the cat away.

"Scat!" she yelled as she rapped her knuckles against the glass. This time it was the cat who was startled as he scrambled awkwardly down the tree. The second he hit the ground, he was gone from sight.

By the time Rosemary had showered and dressed, she'd forgotten about the incident. She headed down the stairs to make breakfast for herself and her own cat, Bubbles. When she reached the kitchen, Rosemary's cat was performing her usual morning ritual of pacing around on the kitchen counter, meowing impatiently for breakfast.

"Bubbles," Rosemary reprimanded, "you know you're not allowed up there. Get down," she said, clapping her hands in the cat's direction. And, as was also the usual morning ritual, Bubbles completely ignored her owner's command, choosing instead to increase the intensity of her meows. "Oh, you must be extra hungry this morning," Rosemary commented as she reached into the cabinet to retrieve the cat's bag of food. Bubbles responded by leaping off the counter and racing to the empty food bowl in the corner of the kitchen. "I'm coming, I'm coming," Rosemary said as she bent down and filled the bowl. As soon as she heard the first piece of dry food hit the bottom of the dish, Bubbles stuck her face in the bowl, causing Rosemary to end up pouring food over the top of her head. This was the way every breakfast began at Rosemary's house.

As Rosemary continued filling the bowl, she happened to glance out the window. Walking along the railing of the deck was the steel-gray cat. Not wanting to disturb Bubbles's breakfast, Rosemary didn't try to scare him away.

After breakfast, Rosemary was washing the dishes in the sink when she looked out the window a second time. "It's you again," she muttered. The cat was still on her deck. He was sitting on her picnic table, watching her through the window. Rosemary raised the window to yell at him but as soon as he saw the window open he was gone in a flash—not, however, before being spotted by Bubbles. Rosemary looked down to see her cat crouched by the sliding glass door, watching the retreating gray cat.

"It's okay," Rosemary whispered. "We chased him away. He won't be back," she added, hoping that it was true.

Unfortunately, the steel-gray cat had other ideas.

"See you tonight," Rosemary said to Bubbles as she headed out the door to work. She and her cat had lived together for only six months. Bubbles was a ten-month-old Siamese. Her daughter and son-in-law had given her to Rosemary. They felt that she was lonely and could benefit from having a pet. Rosemary's husband had died the previous year and her daughter, Nancy, was concerned about her. Getting a dog was discussed, but Nancy remembered that whenever her mother saw a cat, especially a Siamese, she'd comment on how she would someday like to own one. Rosemary's husband had been sick for many years with respiratory problems, so she had dismissed the idea of ever having any kind of pet.

Now, with Bubbles, Rosemary seemed happy. Bubbles was a constant source of amusement, and her ongoing meowing and purring certainly never allowed Rosemary to feel alone.

Rosemary closed her front door and stepped out onto the front porch. It was a beautiful late spring day in Atlanta. She smiled as she looked at her beautiful garden and the lush green grass. Too bad she had to spend so much of the day inside the store where she worked, she thought.

As she approached her old but very dependable car, she noticed two things: one was the beautiful wash and wax job her son-in-law had done for her the previous day. The other thing she noticed was the track of small muddy paw prints that went from the hood, up over the roof, and down the trunk. As she unlocked the door and slid into the driver's seat, Rosemary let out an audible sigh. Something would have to be done about that gray cat.

It wasn't until Rosemary had backed out of the driveway and was about to head down the road that she caught a glimpse of the steel-gray cat sitting on her front porch—watching her. He almost seemed to be smiling at her. Rosemary sounded the car horn and the cat took off. "I'll deal with you later," she said.

At the end of the day, Rosemary returned home to Bubbles. As she opened the door, Bubbles was right there, anxiously waiting. As soon as she saw Rosemary, she began meowing loudly. "Okay,

okay," Rosemary said as she stroked her antsy cat, "I'll get dinner right away." Bubbles seemed about ready to jump out of her skin, almost nervous, Rosemary thought to herself.

On the way to the kitchen, she noticed that several of the magazines on her coffee table were scattered on the floor and the cushions on her couch were also knocked over. "You had fun today, didn't you?" Rosemary commented as she picked up the magazines. It was then that she noticed that one of her plants, the large one that sat in front of the living room window, had also been tipped over. It was not unlike Bubbles to cause some minor destruction during the day, but this was more than usual. She didn't really mind, though, as she enjoyed the cat's sense of adventure. It was nice to have life back in the house.

After she filled Bubbles's food bowl, Rosemary headed back to the living room with a dustpan and broom to sweep up the spilled soil from the overturned pot. While she was kneeling by the window, she smelled a strange odor. She sniffed the soil, but it wasn't coming from there. Then she realized that her knees were feeling damp. She stood up and saw two wet circles on her pants, right at the knees. She felt around and found what was left of a puddle on the floor. It was cat urine. Based on the location, Rosemary figured that Bubbles must have seen the cat outside and gotten upset. That would explain the overturned plant and the puddle on the floor. She assumed Bubbles was trying to mark her territory. So out came some paper towels and disinfectant cleaner. When she was through, Rosemary stepped out of her pants and dropped them right into the washing machine.

In her bedroom, she kicked off her shoes and socks and slipped into a pair of sweatpants. Sitting on the edge of the bed, she rubbed her sore feet for a moment and thought about taking a short nap before making dinner. Bubbles appeared at the door. "Come on over," Rosemary said as she patted the foot of the bed. Bubbles seemed very nervous and began pacing at the doorway. "Well, what's the matter with you? Does that cat outside have you still spooked?"

As she reached down for Bubbles, the cat let out a long, mournful meow and took off down the stairs.

Too tired to chase after her, Rosemary reached over to pull the comforter over herself and felt a wet spot. Putting her fingers to her nose, she sniffed and caught a strong whiff of urine. "Oh, Bubbles, not again," she sighed as she began pulling the linens off the bed. "I guess laundry is on my schedule for this evening."

Later that night, as the laundry tumbled around in the dryer, Rosemary made a mental note to call the veterinarian in the morning. Something had to be done about that gray cat.

Rosemary's instruction from the vet the following morning was to bring Bubbles in for a urinalysis, just in case there was a medical reason why the cat was urinating outside of the litter box. Rosemary dutifully hauled the cat in, who complained during the entire trip there and back. The test results showed no indication of any urinary tract disease. During the exam, though, the vet noted that the cat was in heat and strongly suggested that she be spayed after her heat cycle. Rosemary, who many years ago had had a pet die during just such a surgery, was reluctant. Even after the doctor explained how much veterinary medicine had advanced and how many monitoring devices are used now, Rosemary still wasn't ready.

"That cat outside is going to keep coming around, then," the doctor stated, "and, if Bubbles manages to escape, she'll very likely end up pregnant."

Rosemary shook her head. "She won't ever get out." Just how to get rid of the other cat was another matter, though. "How do I deal with Bubbles tinkling in the house?"

The vet scratched his head, leaned his back against the wall, and folded his arms across his chest. "Mrs. Lockmiller," he began, "the best way to deal with it is to have Bubbles spayed. Other than that, you'll just be cleaning up urine from various locations around the house because the outdoor cat will keep hanging around."

Rosemary looked down at Bubbles. "I need time to think about it," she replied. "Isn't there anything I can do meanwhile?"

Dr. Manning suddenly remembered that this woman had only recently lost her husband. She wasn't ready to face even the remote possibility of losing someone else. "I have someone for you to call," he said. "The receptionist will give you the number of a feline behaviorist." Dr. Manning was hoping that I would be able to convince Rosemary to have Bubbles spayed.

Skeptical about the idea of consulting a feline behaviorist, considering that she'd never even heard of such a thing, Rosemary chose instead to handle the situation herself.

For the next two weeks, she routinely cleaned urine from the carpets, furniture, floors, and walls. Her dustpan and broom were kept out on the table because she'd find herself frequently having to sweep up the broken pieces of knocked over knickknacks and the spilled soil from overturned plants. She kept the blinds closed in the house and faithfully chased the steel-gray cat away every time she saw him. And see him she did. He sat perched in the tree outside of her bedroom window every morning. He serenaded Rosemary and Bubbles every night with a selection of yowls, meows, and cries. He slept on the deck, lounged on Rosemary's car, and helped himself to several unfortunate birds and chipmunks who happened to wander into the yard.

Rosemary's son-in-law had even set humane traps for the cat, but he only managed to snare the same frustrated raccoon time and time again.

Then, for some unexplainable reason, things suddenly quieted down. Bubbles stopped urinating outside of the box and no longer knocked things over inside the house. Two weeks after this whole commotion began, everything went back to normal. Rosemary decided that Bubbles had finally gotten used to seeing that troublesome cat outdoors.

During one visit, Rosemary's daughter remarked how she hadn't seen Bubbles the last few times she'd been over.

"Bubbles has settled down lately," Rosemary said. "I think

she's finally growing out of her wild kitten phase. I also think she has accepted having that ratty old cat in her yard. Thank goodness, because I was at my wit's end with all this tinkling everywhere."

"I didn't think Siamese cats ever settled down," Nancy remarked. "Are you sure Bubbles isn't still upset over that other cat?"

"Romeo doesn't hang around as much as he used to," Rosemary answered.

"*Romeo?*" her daughter said.

"I named him that after he started his nightly serenades under my window."

"Well, I'm just glad this whole incident is over."

"Me too," Rosemary said as she got up from the kitchen table to start preparing dinner. "There is just one thing," she added.

"What is it?"

"This whole episode caused me to gain weight," Rosemary said, and patted her sides. "I guess the stress did it. Even Bubbles put on weight, so maybe she did get traumatized after all. All I know is that we both need to go on diets now."

Rosemary's daughter just shook her head. "I still think you should call that cat lady that the vet told you about. Bubbles is a Siamese cat—she shouldn't get heavy and she shouldn't settle down."

Rosemary thought for a moment. "You know, I really do miss the fact that she doesn't play the way she used to."

Nancy stood up and looked at the clock. "It's not too late. I'll bet the vet's still there." She picked up the phone and called. When she got through to the vet, she asked him whether he felt a visit from the cat shrink was still needed. He answered that if the cat still appeared to be stressed over the outside cat's appearance, it would be a good idea to call the behaviorist. After hanging up the phone, Nancy realized that she'd forgotten to mention Bubbles's weight gain to the vet. Oh well, she thought, her mother can talk to the cat shrink about it.

The next day I received a call from Rosemary's daughter. She told me that her mother was still very skeptical about my services (no surprise there) but that she had finally agreed to it. She described the scenario to me and asked if I could come over right away.

I called Dr. Manning to discuss the case with him because I never agree to see a cat until after a veterinary examination has been done. He felt confident, however, that this was behavioral in origin, triggered by the gray cat's appearance. I agreed to the house call.

Two days later I was sitting in Rosemary Lockmiller's kitchen, listening to her tell the story of how Romeo had spent two weeks relentlessly stalking, wooing, harassing, and peeping in the windows. It sure sounded like a tomcat in pursuit of a female.

"The reason you're here," Rosemary informed me, "is because we feel that Bubbles may still be somewhat upset over the whole episode."

"That's very possible," I replied.

She took me on a tour of the house, pointing out the seemingly endless places where Bubbles had, in Rosemary's words, *tinkled.* Even though she had cleaned the area and Bubbles hadn't gone outside of the box in weeks, the urine odor was still very detectable—overpowering, in fact.

Rosemary must have noticed that I was wrinkling up my nose at the odor. "I know it's bad," she said, "but I just can't get the smell out."

"It's hard to believe that a little Siamese female could create such a vast mess," I commented.

"Oh, she's not so little," Rosemary corrected me.

I looked down at the notes I'd taken on Bubbles's medical history when I'd spoken with Dr. Manning. I'd written the cat's weight at seven pounds. Maybe I'd recorded it wrong. If so, I'd correct it when I saw the cat, I thought.

"Would you like to see Bubbles?" asked Rosemary.

"Yes, I would," I answered, glad for the opportunity to move away from the urine-stained area.

"Normally—before Romeo came into our lives, that is—Bubbles would've been right here to greet you," she said sadly, "but lately she's been sleeping more and tends to stay on my bed during the day."

I followed Rosemary up the stairs and into her bedroom. Just as she had predicted, Bubbles was stretched out on the bed. Something wasn't right, though. I quietly walked toward the bed. I looked at Bubbles and then back at Rosemary, confused. Rosemary just looked blankly back at me. I sat down on the bed and began stroking the cat. Why hadn't anyone told me, I wondered. Then it occurred to me . . . oh my gosh, *they don't know!*

Rosemary spoke suddenly. "I don't know what to do about her diet. She's getting so fat."

I stopped petting Bubbles and looked up at Rosemary. "This cat isn't fat, she's *pregnant.*"

As if relieved that her secret was out, Bubbles rolled over onto her back for a full stretch, exposing her distended belly.

"Pregnant?! Did you say pregnant?" Rosemary laughed. "You're obviously mistaken."

"No, I'm quite certain she's pregnant," I assured her.

"That's impossible. She doesn't go outdoors," Rosemary replied.

"Well, she got out at least once."

"I'm very careful," she defended, obviously losing patience with me. "Bubbles has never been outside."

"She connected with a male cat somehow . . . ," I said, and then a thought occurred to me. The urine odors in the house smelled so strong—as strong as . . . tomcat urine. Maybe I was approaching this the wrong way. Bubbles wasn't getting *out*side; Romeo was getting *in*side. Was that possible? But how? I looked again at Bubbles's big belly and thought, yes, it's not only possible, it's certain.

When I mentioned this to Rosemary, her reaction was . . .

well . . . let's just say she was rather animated. She asked me what kind of crackpot I was to think she wouldn't know if a strange cat was getting into her house.

"Well," I began, undaunted by the fact that I'd just been referred to as a crackpot, "your cat is positively pregnant and I know one thing for sure—she didn't get that way by herself. Now, if you're so sure that she couldn't have gotten outside, then we need to look around for any way that Romeo could've gotten in," I said, convinced I was right that Romeo had, in fact, been in the house, because of the unmistakably strong urine odor.

"Search for yourself," Rosemary stated flatly, then turned toward Bubbles. "How can you be pregnant?"

"Let's begin searching," I said. Rosemary turned and reluctantly followed me out the door.

I began in the basement, checking every window. All was secure down there; besides, Rosemary claimed that she didn't go down there very often and always kept the door locked. So I moved to the main floor. As I examined every window, I could hear Rosemary behind me, reminding me that this was a waste of time because she never opens the windows on the first floor for security reasons. They always remained closed and locked.

I went outside next and checked around the house. Everything looked buttoned up. How did Romeo get in?

I stood in Rosemary's backyard in the blazing summer heat, staring at her house. The answer was somewhere—but where? I walked over to the picnic table to sit under the umbrella. Rosemary was standing by the sliding glass door. "It's too hot out here. I'll be inside if you need me," she said and then quickly disappeared into the house. I imagine that at that point she was pretty darn sure that she wouldn't be paying any money to me for my services.

Seconds after she left, I heard the sound of a sneeze behind me. I turned to find a gray cat standing on the railing of the deck. "Well hello, Romeo," I said. The gray cat blinked in the sun and then delicately walked along the railing, ending up on the far side of the deck. "You don't trust me either, huh?" I asked the cat. He watched me from that position for a while and then turned his attention toward the house. "Romeo, you and I know that I'm not a crackpot," I whispered, "so show me your secret. How are you getting in the house?"

Romeo ignored me and continued looking at the house. His attention then shifted to the large tree next to the deck—the one Rosemary said he had climbed in order to peer in her window. He meowed several times while looking up at the tree. My eyes followed the tree all the way up to the top. There were several upper branches that came so close to the house—specifically that one upstairs window. But if Rosemary said she never opened the windows for security reasons . . . wait, I thought, she said she never opens the *main floor* windows. Squinting in the sun, I strained to look up at the window. There was something odd about the window. Then I saw it—I had the answer. I jumped up from the table, startling Romeo in the process, and ran into the house.

"Rosemary!" I called.

She was seated at the kitchen table. She had apparently been watching me, the crackpot, through the window. "What?" she answered.

I pulled out one of the kitchen chairs and sat across from her. "You said you never open the downstairs windows," I said, "but what about the upstairs ones?"

"Until the weather got too hot in the last couple of weeks, I used to keep several of those windows open. I don't anymore, though, because I have the central air-conditioning on," she answered, looking at me with a puzzled expression. "Why?"

I stood up and headed out of the kitchen, motioning for her to follow. "I know how Romeo got in."

As we ascended the stairs, Rosemary again attempted to take the wind out of my sails. "If you're going to tell me that the cat came through an open window up here, you're crazy. I have screens on all of the windows."

So far that morning, I'd been called crazy and a crackpot. It was shaping up to be another typical day for me. I chose to ignore Rosemary's comment and headed right for the master bedroom instead.

There were two windows in the bedroom. I approached the one closest to the tree. As I began raising the sash, Rosemary again reminded me that there was a screen on this window.

With the window now completely open, it was very clear how Romeo had gotten in. What I'd seen when I was outside looking up at the window was evident now. The lower left portion of the screen was ripped and had been pushed out. When I showed it to Rosemary, she gasped. "You mean that straggly old tomcat really has been in my house?"

I bent closer to examine the ripped screen and pulled several small tufts of gray hairs out. "Yes, he has," I answered, holding up the gray hairs in front of her. "He not only got in the house and mated with your cat, he, not Bubbles, was doing the urine spraying. Of course, when you permanently shut the window a couple of weeks ago, you cut off his entry." Rosemary stood before me with her mouth open. I went on, "Luckily, when you closed the window the last time, Romeo was outside and not inside."

Score one for the crazy crackpot.

One month after my visit to Rosemary's home, Bubbles gave birth to four kittens.

Rosemary called me again when the kittens were three weeks old to say that she hadn't seen Romeo since the week before the kittens were born. She believed that he was gone for good.

If you have an unspayed female cat and you happen to see a steel-gray cat hanging around, watch out. Romeo loves them then leaves them.

Some men just aren't cut out for fatherhood.

○●○

Further Reading

If you are interested in finding specific help for solving a particular behavior problem your cat may be experiencing, please refer to my previous books:

Think Like a Cat How to Raise a Well Adjusted Cat and Not a Sour Puss. New York, NY: Penguin Books, 2000.

Twisted Whiskers: Solving Your Cat's Behavior Problems. Freedom, CA: Crossing Press, 1994.

Psycho Kitty: Understanding Your Cat's Crazy Behavior. Freedom, CA: Crossing Press, 1998.

***

For general-interest reading, here are a few of my favorite books on animals.

Camuti, Dr. Louis J. *All My Patients Are Under the Bed.* New York: Simon & Schuster, 1980.

Caras, Roger A. *A Cat Is Watching.* New York: Simon & Schuster, 1989.

Herriot, James. *James Herriot's Cat Stories.* New York: St. Martin's Press, 1994.

Moussaleff, Masson and Jeffrey McCarthy, and Susan McCarthy. *When Elephants Weep.* New York: Delacorte Press, 1995.

Schoen D.V.M., Allen M., and Pam Proctor. *Love, Medicine and Animal Healing.* New York: Simon & Schuster, 1995.

Resources

Interactive Cat Toys

In addition to increasing your cat's pleasure during play sessions, interactive toys are valuable behavior-modification tools. Here are some recommendations:

Da Bird
Go Cat
3248 Mulliken Road
Charlotte, MI 48813
(517) 543–7519
Best interactive toy ever! Available at pet-supply stores. Your cat will do flips and leaps with this toy, which simulates a bird in flight. Great for cats with high activity levels.

The Kitty Tease
Galkie Company
P. O. Box 20
Harrogate, TN 37752
(800)82–KITTY
Available at many pet-supply stores, or you can order directly from the company.

The Cat Dancer
Cat Dancer Products, Inc.
6145 Green Valley Road
Neenah, WI 54956
(800) 844–6369
www.catdancer.com
Available at pet-supply stores. A favorite among cats, this enticing toy dangles on the end of a wire. There is also a version that can be attached to the wall so that the cat can play when you're not around.

Play-n-Squeak
Our Pet's Virtu Co.
1300 East Street
Fairport Harbor, OH 44077-5573
(800) 565–2695
www.virtupets.com
A sound-generating interactive toy that squeaks like a mouse. They also make another great toy called Play-n-Treat.

The Cat Charmer
Cat Dancer Products, Inc.
(see address above)
Available at pet-supply stores. This snakelike toy works on even the most sedentary felines. Even lazily dragging it along the floor can spark your cat's interest.

Quickdraw McPaw
Claworks
1821 N.W. 65th St.
Seattle, WA 98117
A "peek-a-boo" type toy that's great for timid cats.

Other Products of Interest

Nature's Miracle
Pets 'n People, Inc.
27520 Hawthorne Blvd.
Rolling Hills Estates, CA 90274
(310) 544–7125
Available at pet-supply stores. Neutralizes urine odor and removes stains from carpets, walls, floors, and furniture. Pets 'n People also makes a black light that can be used to determine the exact location of urine stains.

Felix Katnip Tree
Felix Company
3623 Fremont Avenue N.
Seattle, WA 98103
1–800–24–FELIX
Available by ordering directly from the company. Best scratching post ever made!